Real Math
PRACTICE WORKBOOK
Grade 5

Stephen S. Willoughby

•

Carl Bereiter

•

Peter Hilton

•

Joseph H. Rubinstein

•

Joan Moss

•

Jean Pedersen

Columbus, OH

SRAonline.com

Copyright © 2007 by SRA/McGraw-Hill.
All rights reserved. Except as permitted under the United States Copyright Act, no part of this publication may be reproduced or distributed in any form or by any means, or stored in a database or retrieval system, without the prior written permission of the publisher, unless otherwise indicated.

Printed in the United States of America.

Send all inquiries to:
SRA/McGraw-Hill
8787 Orion Place
Columbus, OH 43240-4027

ISBN 0-07-603739-8

3 4 5 6 7 8 9 BCH 12 11 10 09 08 07 06

Table of Contents

Practice Lessons

1.1	Estimating and Measuring	1
1.2	Graphing Height	2
1.3	Place Value	3
1.4	Applying Math	4
1.5	Order and Parentheses	5
1.6	Arithmetic Laws	6
1.7	Adding Multidigit Numbers	7
1.8	Subtracting Multidigit Numbers	8
1.9	Applying Addition and Subtraction	9
1.10	Roman Numerals	10
2.1	Multiplying Multiples and Powers of 10	11
2.2	Multiplying by a One-Digit Number	12
2.3	Multiplying Any Two Whole Numbers	13
2.4	Applying Multiplication	14
2.5	Problem-Solving Applications	15
2.6	Interpreting Remainders	16
2.7	Dividing by a One-Digit Divisor	17
2.8	Exponents	18
2.9	Prime and Composite Numbers	19
2.10	Applications Using Customary Measurement	20
2.11	Temperature	21
3.1	Decimals and Money	22
3.2	Place Value and Decimals	23
3.3	Comparing and Ordering Decimals	24
3.4	Adding and Subtracting Decimals	25
3.5	Applying Math	26
3.6	Multiplying and Dividing Decimals	27
3.7	Metric Units	28
3.8	Choosing Appropriate Metric Measures	29
3.9	Multiplying Decimals by Whole Numbers	30
3.10	Rounding and Approximating Numbers	31
3.11	Approximation Applications	32
3.12	Understanding Decimal Division Problems	33
3.13	Interpreting Quotients and Remainders	34
3.14	Decimals and Multiples of 10	35
3.15	Applying Decimals	36
4.1	Using Your Calculator	37
4.2	Using Number Patterns to Predict	38
4.3	Repeated Operations: Savings Plans	39
4.4	Function Machines	40
4.5	Multiplication Function Rules	41
4.6	Finding Function Rules	42
4.7	Subtraction Rules and Negative Numbers	43
4.8	Adding and Subtracting Integers	44
4.9	Multiplying and Dividing Integers	45
4.10	Patterns	46
5.1	Coordinates	47
5.2	Functions and Ordered Pairs	48
5.3	Composite Functions	49
5.4	Graphing in Four Quadrants	50
5.5	Making and Using Graphs	51
5.6	Inverse Functions	52
5.7	Inverse of a Composite Function	53
5.8	Using Composite Functions	54
5.9	Temperature Conversions	55
5.10	Standard Notation for Functions	56
5.11	Composite Functions in Standard Notation	57
5.12	Linear Equations	58
6.1	Fractions of a Whole	59
6.2	Fractions of Fractions	60
6.3	Decimal Equivalents of Fractions	61
6.4	Equivalent Fractions	62
6.5	Fractions with the Same Denominator	63
6.6	Practice with Fractions	64
6.7	Comparing Fractions	65
6.8	Counting Possible Outcomes	66
6.9	Probability and Fractions	67
6.10	Adding Fractions	68
6.11	Subtracting Fractions	69
6.12	Applying Fractions	70
7.1	Mixed Numbers and Improper Fractions	71
7.2	Multiplying Mixed Numbers	72
7.3	Adding Mixed Numbers	73
7.4	Subtracting Mixed Numbers	74

Real Math • Grade 5 • *Practice*

Table of Contents

Practice Lessons

7.5	Addition and Subtraction Applications	75
7.6	Dividing Fractions	76
7.7	Fractions and Decimals	77
7.8	Decimal Equivalents of Rational Numbers	78
7.9	Using Mixed Numbers	79
8.1	Averages	80
8.2	Mean, Median, Mode, and Range	81
8.3	Interpreting Averages	82
8.4	Ratios and Rates	83
8.5	Comparing Ratios	84
8.6	Using Approximate Quotients	85
8.7	Approximating Quotients	86
8.8	Dividing by a Two-Digit Number	87
8.9	Practice with Division	88
8.10	Dividing by a Three-Digit Number	89
8.11	Batting Averages and Other Division Applications	90
8.12	Average Heights	91
8.13	Using Rates to Make Predictions	92
8.14	Population Density	93
8.15	Using Ratios	94
9.1	Angles	95
9.2	Measuring Angles	96
9.3	Angles and Sides of a Triangle	97
9.4	Drawing Triangles	98
9.5	Congruence and Similarity	99
9.6	Corresponding Parts of Triangles	100
9.7	Using Corresponding Parts of a Triangle	101
9.8	Scale Drawings	102
9.9	Using a Map Scale	103
9.10	Perpendicular and Parallel Lines and Quadrilaterals	104
9.11	Parallelograms	105
9.12	Exploring Some Properties of Polygons I	106
9.13	Exploring Some Properties of Polygons II	107
10.1	Circles: Finding Circumference	108
10.2	Area of Parallelograms	109
10.3	Area of Triangles	110
10.4	Area of a Circle	111
10.5	Area of Irregular Figures	112
10.6	Rotation, Translation, and Reflection	113
10.7	Symmetry	114
10.8	Paper Folding	115
10.9	Making a Flexagon	116
10.10	Space Figures	117
10.11	Building Deltahedra	118
10.12	Surface Area	119
10.13	Volume	120
11.1	Approximating Products of Decimals	121
11.2	Multiplying Two Decimals	122
11.3	Percent and Fraction Benchmarks	123
11.4	Computing Percent Discounts	124
11.5	Computing Interest	125
11.6	Percents Greater than 100%	126
11.7	Probability and Percent	127
11.8	Simplifying Decimal Division	128
11.9	Dividing Two Decimals	129
12.1	Estimating Length	130
12.2	Estimating Angles and Distances	131
12.3	Applying Customary Measures	132
12.4	Converting Measures	133
12.5	Measuring Time	134
12.6	Measuring Circles and Angles	135
12.7	Pictographs and Data Collection	136
12.8	Making Circle Graphs	137
12.9	Creating and Using Graphs	138
12.10	Making Line Graphs	139
12.11	Interpreting Graphs	140

Masters 141

LESSON 1.1

Name _____ Date _____

Estimating and Measuring

Mallory is doing the following measuring activity. First, she estimates the length from a person's shoulder to his or her fingertips and writes it on her table. Next, she measures the actual length and records it. Then, she finds the difference between the estimate and the measurement and records it.

Complete this measuring activity in a small group. Then answer the following questions.

Estimate and then measure the lengths of your group members' arms. Record the results in the table below. Also find the differences between your estimates and measurements. Make a note of whether the estimate or the measurement was greater.

Name	Estimate (centimeters)	Measurement (centimeters)	Difference (centimeters)

❶ Did your estimates get better after the first few? Explain.

_____

❷ Do any students have the same lengths from their shoulders to their fingertips? If so, were the students the same height?

_____

❸ What is the average of all the lengths? _____

Real Math • Grade 5 • *Practice*   Chapter 1 • *Whole Numbers Refresher*

LESSON 1.2

Name _____ Date _____

Graphing Height

Complete the following exercises.

1 Measure the heights of three friends or family members. Then have one of them measure your height. Record your measurements on a piece of paper.

2 Make a bar graph. Put the names in order from the shortest height measurement to the tallest.

Count on or back. Write the missing numbers.

3 26, 27, _____, _____, _____, 31, 32, _____

4 49, 48, _____, _____, 45, _____, _____, 42

5 _____, 76, _____, _____, 79, _____, 81, 82

6 _____, _____, 78, 77, 76, 75, _____, _____

7 _____, 21, _____, _____, _____, 17, 16, 15

8 _____; _____; _____; 997; 998; _____; 1,000

9 343, 342, _____, _____, 339, _____, 337, _____

10 617, 618, _____, _____, 621, 622, _____, _____

11 105, _____, _____, _____, 101, 100, _____

12 _____, _____, _____, 809, 808, _____, 806

Answer the following questions.

13 Jeffrey and Katrina made cookies for the bake sale. Jeffrey made 72 cookies, and Katrina made 66 cookies. Jeffrey wants to find out the total number of cookies they both made. What operation should Jeffrey use? How many cookies did they make altogether?

_____

14 When Shatonna was 9 years old, her allowance was $1 per week. When she was 10, her allowance was $2 per week. At 11 years old, it was $4 per week. When Shatonna was 12, it was $7 per week. If this pattern continues, what will Shatonna's weekly allowance be when she is 14? Describe the pattern.

_____

2 Chapter 1 • *Whole Numbers Refresher*

LESSON 1.3

Place Value

Write the following numbers in standard form. An example has been done for you. Then answer the two questions.

8,000 + 200 + 50 + 4 = 8,254

1. 7,000 + 300 + 90 + 6 = _____
2. 10,000 + 2,000 + 300 + 40 + 5 = _____
3. 1,000 + 800 + 30 + 9 = _____
4. 9,000 + 800 + 50 + 3 = _____
5. 20,000 + 400 + 6 = _____
6. 200,000 + 7,000 + 600 + 40 + 2 = _____
7. 40,000 + 7 = _____
8. 300,000 + 100 + 8 = _____
9. How many digits are in the number five hundred six million? _____
10. If a whole number has eight digits, what is its greatest place value? _____

Answer the following questions using the table.

Surface Areas of the Great Lakes

Lake	Surface Area (km²)	Lake	Surface Area (km²)
Erie	25,657	Ontario	18,960
Huron	59,500	Superior	82,100
Michigan	57,750		

11. Which lake has greater area—Huron or Superior?

_____

12. Is the total area of the two smallest lakes added together more or less than the area of the largest lake? Explain.

_____

13. List the lakes in order from least to greatest area.

_____

14. Is the sum of the areas of the three smallest lakes greater than or less than the sum of the areas of the two largest lakes? Show your work.

_____

_____

Real Math • Grade 5 • *Practice*   Chapter 1 • *Whole Numbers Refresher* **3**

Lesson 1.4 — Applying Math

Name _____ Date _____

Answer the following questions. Show your work.

1. Sydney bought 5 pencils for 93¢. About how much did she pay for each pencil? _____

2. Pentown is 10 miles east of Centerville. Inkville is 5 miles west of Centerville. Markerport is 8 miles west of Centerville. Draw a map to help answer these questions.

   a. How many miles apart are Pentown and Inkville? _____

   b. How many miles apart are Markerport and Pentown? _____

   c. How many miles apart are Markerport and Inkville? _____

3. Saburo and 2 friends bought a box of 6 muffins. If they share the muffins equally, how many will each person get? _____

4. Oscar had $25 when he went to an amusement park. The ticket cost him $9. He spent $3 to play a video game, $5 for food, and $2 for a poster. How much money did Oscar spend altogether? _____

5. In his garden, Cameron wants to have 4 rows of tomato plants with 6 plants in each row. He can buy 2 tomato plants for $1.

   a. How much will he pay for the tomato plants? _____

   b. If each tomato plant in Cameron's garden produced 4 tomatoes, how many dozen tomatoes would that be? _____

Answer the following questions by thinking about each situation.

6. Think about a stack of cubes that looks like the one in the figure, with 4 cubes along each edge. How many cubes are in the entire stack? _____

7. Suppose you painted the outside of this stack. How many of the cubes would not be painted at all? _____

   How many would be painted on only one face? two faces? three faces? four? five? six? no faces? _____

   _____

   Do the numbers add up to 64? _____

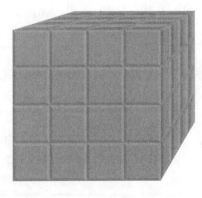

4 Chapter 1 • *Whole Numbers Refresher*

Real Math • Grade 5 • *Practice*

LESSON 1.5
Order and Parentheses

Solve for *n*.

① $(6 + 3) \times 2 = n$ _____

② $3 + (6 \times 2) = n$ _____

③ $(20 \div 5) \times 4 = n$ _____

④ $20 \div (5 \times 4) = n$ _____

⑤ $(4 + 3) \times (5 - 1) = n$ _____

⑥ $30 \div (8 - 2 - 1) = n$ _____

⑦ $(9 + 7) \div (6 - 4) = n$ _____

⑧ $(3 \times 8) \div (9 - 5) = n$ _____

For each problem, tell how many different answers you can get by putting the parentheses in different places and solving. The first two are done for you.

$4 + 6 + 8 = n$
$(4 + 6) + 8 = 18$
$4 + (6 + 8) = 18$
There is only one possible answer: $n = 18$.

$7 - 4 - 3 = n$
$(7 - 4) - 3 = 0$
$7 - (4 - 3) = 6$
There are two possible answers: $n = 0, 6$.

⑨ $36 \div 6 \div 2 = n$ _____

⑩ $18 + 4 \times 2 = n$ _____

⑪ $8 + 7 + 2 = n$ _____

⑫ $7 \times 3 \times 2 = n$ _____

⑬ $10 + 5 \times 2 = n$ _____

⑭ $9 - 6 - 1 = n$ _____

⑮ $4 \times 10 \div 2 = n$ _____

⑯ $3 \times 5 \times 4 = n$ _____

LESSON 1.6

Name _____ Date _____

Arithmetic Laws

Complete each exercise. Use shortcuts when you can.

1. $52 + 93 =$ _____
2. $93 + 52 =$ _____
3. $4 \times 9 =$ _____
4. $9 \times 4 =$ _____
5. $4 + (96 + 75) =$ _____
6. $(96 + 75) + 4 =$ _____
7. $9{,}354 \times 0 =$ _____
8. $258 \times 1 =$ _____
9. $258 + 0 =$ _____
10. $9 \times 49 \times 0 \times 716 =$ _____
11. $100 \times 93 =$ _____
12. $(46 \times 93) + (54 \times 93) =$ _____
13. $(38 \times 76) + (62 \times 76) =$ _____
14. $8 \times 7 \times 6 \times 5 \times 4 \times 3 \times 0 =$ _____
15. $7 \times (300 + 4) =$ _____
16. $342 \times 85 \times 1 \times 29 \times 0 =$ _____
17. $(28 \times 55) + (72 \times 55) =$ _____
18. $0 + 1 + 3 + 5 + 7 + 9 =$ _____

Solve the following problems.

19. If tomorrow is April 1, how many days are there until October 31? Explain how you can get your answer without adding the number of days in each of the seven months.

    _____
    _____
    _____

20. Use the distributive law to show how many days there are in June, July, August, September, and October.

    _____

6 Chapter 1 • *Whole Numbers Refresher*   Real Math • Grade 5 • *Practice*

LESSON 1.7

Name _____ Date _____

Adding Multidigit Numbers

Find the sum. Use shortcuts when you can.

1. 34 + 56 = _____
2. 19 + 63 = _____
3. 167 + 42 = _____
4. 4,973 + 535 = _____
5. 7,921 + 46 = _____
6. 7,142 + 374 = _____
7. 435 + 9,765 = _____
8. 1,296 + 5,742 = _____
9. 3,443 + 5,225 = _____
10. 2,040 + 2,040 = _____
11. 371 + 600 = _____
12. 492 + 365 = _____
13. 3,090 + 10 = _____
14. 2,765 + 3,829 = _____
15. 854 + 9,027 = _____
16. 375 + 425 = _____
17. 200 + 300 + 450 = _____
18. 3,974 + 128 + 97 = _____
19. 201 + 397 + 863 + 1,124 = _____
20. 1,000 + 200 + 3,000 + 6,500 = _____

In each exercise, two of the answers do not make sense and one is correct. Ring the correct answer. Think about your methods of finding answers. Which methods work best?

21. 829 + 231
    a. 1,060
    b. 1,270
    c. 920

22. 5,400 + 230
    a. 7,730
    b. 5,630
    c. 5,240

23. 8,497 + 23
    a. 8,690
    b. 84,390
    c. 8,520

24. 21,121 + 33,313
    a. 54,434
    b. 44,444
    c. 55,555

Solve the following problems.

25. A total of 20,877 people attended 4 baseball games. The attendance for the first game was 5,675. The attendance for the second game was 4,810. The third game had 5,121 in attendance. How many people attended the fourth game? Which game had the largest attendance?

    _____

26. What is the greatest sum you can get by adding two four-digit numbers? What is the least sum you can get by adding two four-digit numbers? Explain.

    _____

    _____

    _____

Real Math • Grade 5 • *Practice* Chapter 1 • Whole Numbers Refresher **7**

LESSON 1.8

Name _____ Date _____

Subtracting Multidigit Numbers

Find the difference. Use shortcuts when you can.

①   84
   − 56

②   99
   − 63

③   167
   − 42

④ 7921
  − 46

⑤ 7142
  − 374

⑥ 9435
  − 765

⑦ 8443
 − 5225

⑧ 2040
 − 2040

⑨ 671
 − 600

Add or subtract. Watch the signs.

⑩ 3,090 + 110 = _____
⑪ 3,090 − 110 = _____
⑫ 854 + 146 = _____

⑬ 3,000 − 20 = _____
⑭ 245 + 115 = _____
⑮ 4,004 − 1,100 = _____

⑯ 675 + 325 = _____
⑰ 827 + 173 = _____
⑱ 1,675 − 675 = _____

⑲ 2,000 − 121 = _____
⑳ 3,768 + 1,232 = _____
㉑ 8,763 − 100 = _____

㉒ 5,438 + 562 = _____
㉓ 1,492 − 433 = _____
㉔ 2,441 − 2,441 = _____

Solve these problems.

㉕ The total area of the three smallest states in the United States—Connecticut, Delaware, and Rhode Island—is 9,034 square miles. If the area of Rhode Island is 1,545 square miles and the area of Delaware is 1,945 square miles, what is the area of Connecticut?

_____

㉖ In the subtraction exercise below, each letter stands for a digit from 0–9. Find the digits.

   A34B
 − 12D5
   1C68

A = _____   B = _____   C = _____   D = _____

8 Chapter 1 • *Whole Numbers Refresher*   Real Math • Grade 5 • *Practice*

Lesson 1.9

Name _____ **Date** _____

Applying Addition and Subtraction

Add or subtract. Watch the signs.

❶ 820
   − 500

❷ 59
   − 42

❸ 418
   + 132

❹ 480
   + 240

❺ 6000
   − 187

❻ 922
   + 52

❼ 9999
   − 9815

❽ 2000
   + 1944

Choose the best approximation.

❾ The Treaty of Versailles was signed in 1919, officially ending World War I. About how many years ago was that?
   a. 90   **b.** 175   **c.** 350

❿ The population of the United States in 2004 was about 293,633,000. The area of the United States is about 3,717,796 square miles. About how many people were there per square mile in 2004?
   a. 150   **b.** 80   **c.** 20

Use the map to answer these questions.

⓫ What is the shortest distance from Classic City to Sound City if you go through

   a. Country Junction? _____

   b. Rapville? _____

   c. Rocktown? _____

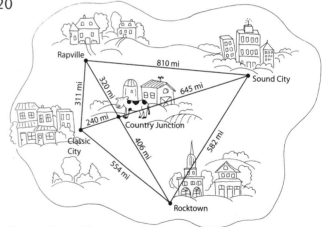

⓬ Mr. Roberts thinks that the trip from Classic City to Rapville to Country Junction to Classic City is longer than the trip from Classic City to Country Junction to Rocktown to Classic City.

   a. Is Mr. Roberts correct? _____

   b. How many more miles are there in the longer trip? _____

Real Math • Grade 5 • *Practice*   Chapter 1 • **Whole Numbers Refresher** 9

Lesson 1.10 — Roman Numerals

Write the year when the following events occurred in Roman numerals.

1. The Louisiana Purchase _____
2. Alaska becomes the 49th state _____
3. The Battle of New Orleans _____
4. Eli Whitney invents the cotton gin _____
5. Jamestown is founded _____

Complete the exercises below. Write the answers in Roman numerals.

6. MCC − MLVII = _____
7. DCC − CD = _____
8. DCC + CCC = _____
9. M − I = _____
10. LXXIX + XXVII = _____
11. LXX + LVII = _____
12. D − LIV = _____
13. CCCXXV + CCCXXV = _____
14. CMLX + MDX = _____
15. LX − XXI = _____

Answer the following questions.

16. Construction of the Empire State Building in New York City began in 1930. How is this year written in Roman numerals on the cornerstone?

17. Sequences of poems, such as Shakespeare's sonnets, are generally numbered using Roman numerals. If you wanted to read Shakespeare's 98th sonnet, which Roman numeral would you look for in the book?

LESSON 2.1

Name _____ Date _____

Multiplying Multiples and Powers of 10

Find each product.

1. $7 \times 10 =$ _____
2. $7 \times 100 =$ _____
3. $7 \times 1{,}000 =$ _____
4. $10 \times 23 =$ _____
5. $23 \times 100 =$ _____
6. $1{,}000 \times 23 =$ _____
7. $58 \times 1{,}000 =$ _____
8. $58 \times 10 =$ _____
9. $58 \times 100 =$ _____
10. $29 \times 10 =$ _____
11. $29 \times 100 =$ _____
12. $29 \times 1{,}000 =$ _____
13. $50 \times 60 =$ _____
14. $500 \times 600 =$ _____
15. $50 \times 60{,}000 =$ _____
16. $3{,}000 \times 5{,}000 =$ _____
17. $300 \times 50{,}000 =$ _____
18. $30{,}000 \times 5{,}000 =$ _____

Answer the following questions.

19. There are 100 stamps in a standard roll. How many stamps are in

    a. 4 rolls? _____   c. 43 rolls? _____

    b. 20 rolls? _____   d. 365 rolls? _____

20. There are 500 sheets of paper in 1 ream. How many sheets of paper are in

    a. 7 reams? _____   c. 70 reams? _____

    b. 10 reams? _____   d. 700 reams? _____

21. Pablo practices the guitar 30 minutes per day on school days. On the weekends, he practices for 1 hour per day. At this rate, for how many minutes does he practice in 2 weeks? Is this greater than or less than 10 hours? Explain.

    _____

    _____

    _____

Real Math • Grade 5 • *Practice*   Chapter 2 • *Multiplication and Division Refresher*

LESSON 2.2

Name _____ Date _____

Multiplying by a One-Digit Number

Find the number of tiles in each hallway from the information given below.

1. 8 tiles wide and 9 tiles long _____
2. 9 tiles wide and 80 tiles long _____
3. 7 tiles wide and 11 tiles long _____
4. 70 tiles wide and 11 tiles long _____
5. 9 tiles wide and 700 tiles long _____
6. 8 tiles wide and 24 tiles long _____

Multiply. Check to see whether your answers make sense.

7. 321 × 3
8. 123 × 3
9. 400 × 8
10. 800 × 4

11. 87 × 6
12. 67 × 8
13. 68 × 7
14. 78 × 6

15. 59 × 4
16. 509 × 4
17. 432 × 1
18. 99 × 0

Solve these problems.

19. Marian has a T-shirt booth at the mall. She calculates how much it costs her to run her booth. She has 3 full-time workers. Each works 9 hours a day, 4 days a week. She pays them $11 per hour. She pays $48 per day to rent the booth. Her other expenses are $75 per week.

    a. What does Marian pay her workers each week? _____
    b. What is her weekly rent? _____
    c. What does it cost Marian to run her booth for 1 week? _____

20. Marian sold 347 T-shirts last week. She earns $4 for each shirt sold. Compare her expenses with her earnings. Did she make more than she spent? Explain.

    _____
    _____

12 Chapter 2 • *Multiplication and Division Refresher*   Real Math • Grade 5 • *Practice*

LESSON 2.3

Name _____ Date _____

Multiplying Any Two Whole Numbers

Multiply. Use shortcuts when you can.

1. 12 × 12 = _____
2. 24 × 12 = _____
3. 24 × 24 = _____
4. 35 × 6 = _____
5. 70 × 6 = _____
6. 72 × 6 = _____
7. 610 × 8 = _____
8. 610 × 9 = _____
9. 611 × 9 = _____
10. 15 × 15 = _____
11. 150 × 15 = _____
12. 150 × 30 = _____
13. 246 × 6 = _____
14. 246 × 60 = _____
15. 246 × 30 = _____

Solve the following problems. Show your work.

16. Acme Concrete Company built a new parking garage. They made it large enough to hold 80 cars on each of 6 levels.

    a. How many cars can the parking garage hold if it is entirely full? _____

    b. The top level of the garage is closed on snowy days. What is the capacity of the garage on those days? _____

    c. The garage charges $4 per hour for cars to park. If the garage is open 10 hours a day and is half full the entire time, how much money will the garage collect? _____

17. Every year the garage repaints the yellow stripes that separate the parking spaces. It takes 5 painters working 8 hours a day for 4 days to finish the job. They get paid $13.50 per hour plus free parking for each day they work. How much does the paint crew earn for their work? _____

18. The yellow paint used to paint the stripes cost $33 per bucket. The paint crew uses 26 buckets of paint and 12 paintbrushes for the job.

    a. How much does the paint cost? _____

    b. How much do the brushes cost? _____

LESSON 2.4

Name _____ Date _____

Applying Multiplication

Use the table to answer the following questions. The data below tells about how many calories a 100-pound person burns in 1 hour doing each activity listed.

Calories Burned Per Hour

Activity	Calories Per Hour	Activity	Calories Per Hour
Beach volleyball	384	Skateboarding	240
Bowling	144	Sleeping	30
Jumping rope	480	Soccer	336
Mountain biking	408	Tai chi	192

1. Marla played 3 hours of beach volleyball on Saturday and 4 hours of beach volleyball on Sunday. How many calories did she burn on those two days? _____

2. Jorge burned 4,080 calories doing one of the activities for 10 hours last week. Which activity did he do? How do you know?

   _____

3. Elena jumped rope for 1 hour. After a short break for a snack, she went skateboarding for 2 hours. How many calories did Elena burn on those two activities? _____

4. During one weekend, Fasil played soccer for 6 hours. How many calories did he burn? _____

5. Which burns more calories—3 hours of bowling or 2 hours of tai chi? How much more? _____

6. Dillon went bowling for 2 hours. Then he rode his mountain bike for 2 hours. Ellie jumped rope for 1 hour and then went skateboarding for 3 hours. Who burned more calories? How many more? Explain.

   _____

7. Sophia's goal was to burn 1,000 calories. She played soccer for 2 hours with her friends, took a 1-hour nap, and then jumped rope for a half hour. Did she meet her goal? Explain.

   _____

LESSON 2.5

Name _____ Date _____

Problem-Solving Applications

Read each problem. Think carefully about what you need to do to solve it, and then answer the questions.

❶ Each side of Pentagon Park measures 320 yards.

a. David walks around the park 3 times each day. How many yards does David walk altogether? _____

b. David jogs around Pentagon Park 4 times each week. Each run includes 5 full loops of the park. What is the total distance David runs in a week? _____

c. A mile is 1,760 yards. Estimate about how many times you must walk around Pentagon Park to cover 3 miles. Explain.

_____

_____

❷ Nancy is printing 48 copies of a family-reunion photograph to send to the relatives who attended. The cost per copy is 79¢.

a. How much money will Nancy need to buy enough copies for everyone who attended the reunion? _____

b. Nancy gets 5 additional copies enlarged and framed. The cost is $2 per enlargement and $14 per framed photograph. What is the cost for this part of the order? _____

c. Nancy pays for the total order with 6 $20 bills. How much change should she receive? _____

❸ Jared takes the bus into New York City for the day. His expenses include a theater ticket, lunch, snacks, souvenirs, and a gift for his parents.

a. Jared got a half-price ticket to a Broadway show. The regular ticket price is $78. How much did Jared spend, including the $5 service charge? _____

b. Lunch at the City Café cost $14. The tax was $1.12. He left a $2.50 tip. How much did he spend for lunch? _____

c. After the show Jared bought a souvenir mug for $8.50, a T-shirt for $22.95, and a CD for $15. All prices included tax. How much did he spend after the show? How much did Jared spend on his day in the city altogether? _____

Real Math • Grade 5 • *Practice*   Chapter 2 • *Multiplication and Division Refresher* **15**

Name _____ **Date** _____

Interpreting Remainders

Find the quotient, and write the remainder if there is one.

1. $4\overline{)31}$
2. $4\overline{)34}$
3. $6\overline{)40}$
4. $8\overline{)40}$
5. $6\overline{)49}$
6. $8\overline{)46}$
7. $1\overline{)17}$
8. $8\overline{)76}$

Solve each problem. Think about how to deal with the remainder.

9. The Van Go Company is renting vans to Oneida School for a field trip. Each van can hold 8 students and 1 driver. How many vans are needed to carry 60 fifth graders from Oneida School to the nature reserve? Explain.

   _____
   _____

10. Chef Jeff bought a basket of apples from a fruit stand at his neighborhood farmers' market. The basket held 58 apples. His favorite pie recipe calls for 6 apples.

    a. How many pies can Chef Jeff make? _____

    b. How many apples will be left over after he makes as many pies as possible with 58 apples? _____

11. A machine at the Eagle Eye Golf Course fills buckets with golf balls for people to use for practice. Toby fills the machine with golf balls so the machine can fill the buckets. Each bucket can hold up to 60 balls. Toby has 8 cases of 400 balls to put into the empty machine.

    a. How many buckets can the machine fill after Toby empties all 8 cases into the machine if no other balls are added? _____

    b. If any balls are left over, Toby usually practices with them after work. How many balls will Toby get to hit? _____

12. Hot dog buns usually come in bags of 12. Hamburger buns come in bags of 8. The cafeteria plans to serve 250 hot dogs and 250 hamburgers at the class picnic.

    a. How many bags of hot dog buns are needed? _____

    b. How many bags of hamburger buns are needed? _____

16 Chapter 2 • *Multiplication and Division Refresher* Real Math • Grade 5 • *Practice*

LESSON 2.7

Name _____ Date _____

Dividing by a One-Digit Divisor

Find the quotient, and write the remainder if there is one.

1 2)96   **2** 3)96   **3** 4)96

4 5)79   **5** 5)792   **6** 4)255

7 4)257   **8** 8)371   **9** 8)765

Find each missing digit.

10 375 ÷ 5 = 7☐

11 1☐6 ÷ 6 = 31

12 ☐76 ÷ 4 = 44

13 360 ÷ 10 = 3☐

14 2☐6 ÷ 6 = 41

15 332 ÷ 5 = 66R☐

16 2,4☐2 ÷ 8 = 306R4

17 556 ÷ 7 = 79R☐

Solve the following problems.

18 Mr. Link lets 6 students go to the computer lab every hour. How many hours will it take for all 28 students in his class to get a turn?

_____

19 The students in Ms. Fried's class are forming groups to do skits for Black History Month. Each group can have no more than 4 members. The class has 34 students. What is the least number of groups there could be?

_____

20 All the fifth and sixth graders at Bach School attended a dance performance in the auditorium. The students completely filled 8 rows, and 5 students sat in the ninth row with the teachers. Each row has 22 seats. How many students attended the performance? Explain.

_____

Real Math • Grade 5 • *Practice*   Chapter 2 • *Multiplication and Division Refresher*

LESSON 2.8

Name _____ Date _____

Exponents

Write each product using exponents.

1 $6 \times 6 \times 6 \times 6 \times 6 \times 6 \times 6 \times 6 =$ _____

2 $9 \times 9 \times 9 \times 9 \times 9 \times 9 =$ _____

Evaluate these by using multiplication or the exponent key on a calculator.

3 $3^4 =$ _____   **4** $5^6 =$ _____   **5** $6^5 =$ _____

6 $10^4 =$ _____   **7** $4^7 =$ _____   **8** $9^5 =$ _____

9 $1^{10} =$ _____   **10** $5^7 =$ _____   **11** $7^{10} =$ _____

Use the facts in the table to solve the exercises below. Do *not* use your calculator. Write your answers in standard form (without exponents).

$6^1 = 6$	$6^5 = 7,776$	$6^9 = 10,077,696$
$6^2 = 36$	$6^6 = 46,656$	$6^{10} = 60,466,176$
$6^3 = 216$	$6^7 = 279,936$	$6^{11} = 362,797,056$
$6^4 = 1,296$	$6^8 = 1,679,616$	$6^{12} = 2,176,782,336$

12 $6^3 \times 6^2 =$ _____   **15** $6^9 \times 6^3 =$ _____

13 $6^5 \times 6^4 =$ _____   **16** $279,936 \times 7,776 =$ _____

14 $6^5 \times 6^5 =$ _____   **17** $46,656 \times 46,656 =$ _____

Solve the following problem.

18 Belinda is saving money for her summer vacation. She puts away $2 each week for 20 weeks. Marcus is also saving for his vacation, but he puts away 2 cents each week and doubles the amount each week for 14 weeks. Whose plan will result in the greater amount of savings? Which plan would you rather follow? Explain.

_____

_____

_____

LESSON 2.9

Prime and Composite Numbers

Write all the factors for the following numbers.

1. 8 _____
2. 9 _____
3. 15 _____
4. 40 _____
5. 43 _____
6. 200 _____

Decide whether each number is prime or composite. If it is prime, write *P*. If it is composite, write the number in factored form as a product of prime numbers. Use exponents when there are two or more identical prime factors.

7. 13 _____
8. 14 _____
9. 27 _____
10. 48 _____
11. 49 _____
12. 56 _____
13. 68 _____
14. 71 _____
15. 102 _____
16. 121 _____
17. 168 _____
18. 192 _____
19. 250 _____
20. 450 _____
21. 4,500 _____

Solve the following problems.

22. Jamilla's mother is 28 years older than Jamilla. Both their ages are prime numbers. The sum of their ages is 54. How old is Jamilla? How old is her mother? _____

23. Two prime numbers have a sum of 54 and a difference of 20. What are the numbers? _____

24. The ages of the three children in the Suvari family are prime numbers. The oldest is 10 years older than the youngest. Altogether, their ages total 35 years. How old are the three Suvari children? _____

Name _____ **Date** _____

Applications Using Customary Measurement

1 ton = 2,000 pounds	1 foot = 12 inches	1 hour = 60 minutes
1 pound = 16 ounces	1 mile = 5,280 feet	1 minute = 60 seconds

Each statement has a wrong unit of measure. Write the more sensible unit.

❶ We rode in a taxi that was about 11 inches long. _____

❷ The big box of cereal weighs about 1 ounce. _____

❸ Mr. Ito held his breath for 30 days. _____

❹ The airport is located 8 feet south of the city. _____

Answer the following questions.

❺ How many ounces are in 6 tons? _____

❻ How many pounds are in 768 ounces? _____

❼ The ceiling in the library is 156 inches from the floor. How many feet high is it?
_____

❽ The distance between Galveston and Houston, Texas, is about 264,000 feet. How far is this distance in miles? _____

❾ As of 2004, the world record for the marathon was held by Kenyan runner Paul Tergat. His record time was 2 hours, 4 minutes, and 55 seconds. What is that time in seconds only?
_____

❿ Darcy has 200 minutes on her calling card. She talks to her brother one night for 1 hour and 45 minutes. The next night she chats with one friend for 20 minutes and another friend for 1 hour and 5 minutes. Does she have enough time left on her card to make a half hour call to her best friend? Explain. _____

20 Chapter 2 • *Multiplication and Division Refresher*

LESSON 2.11

Name _____ Date _____

Temperature

Use this table to answer the questions that follow. The table shows mean (or average) temperatures (Fahrenheit) for four American cities.

	JAN	FEB	MAR	APR	MAY	JUN	JUL	AUG	SEP	OCT	NOV	DEC
Albany, NY	22	25	35	47	58	66	71	69	61	49	39	28
Bismarck, ND	10	18	30	43	56	65	70	69	58	45	28	15
Honolulu, HI	73	73	74	76	77	80	81	82	82	80	78	75
Tampa, FL	61	63	67	72	78	82	83	83	82	76	69	63

1 a. What is the coldest month of the year in Albany? _____

  b. What is the coldest month of the year in Bismarck? _____

  c. What is the coldest month of the year in Honolulu? _____

  d. What is the coldest month of the year in Tampa? _____

2 a. Which city has the greatest difference between its highest and lowest monthly temperatures? _____

  b. What is that difference? _____

  c. Which months did you compare? _____

3 During which consecutive months is Tampa's average temperature about the same? Explain.

_____

4 How would you summarize the climate in Honolulu?

_____

5 Between which two consecutive months does Bismarck experience the greatest change in average temperature? What is the diffference?

_____

6 During which month would you expect a spring thaw in Albany? Explain.

_____

7 Which two cities have the most similar climates? Explain.

_____

_____

Real Math • Grade 5 • *Practice*   Chapter 2 • *Multiplication and Division Refresher*

Name _____ Date _____

Decimals and Money

Use <, >, or = to complete each statement.

1. $9.07 ◯ $7.09
2. $4,015.29 ◯ $4,051.19
3. $204.02 ◯ $202.04
4. $975.00 ◯ $975
5. $63.48 ◯ $634.80
6. $8.09 ◯ $8.90
7. $0.97 ◯ 97¢
8. $47.99 ◯ $479
9. $92.00 ◯ $9.20
10. $0.02 ◯ $2.00
11. $3.00 ◯ $3
12. $270.27 ◯ $27.27

Add or subtract. Watch the signs.

13. $92.57
    − 67.85

14. $74.92
    + 25.59

15. $1.15
    + 2.75

16. $294.15
    + 379.62

17. $273.60
    + 326.40

18. $30.00
    − 29.37

Solve these problems.

19. William had $50. He bought a video game for $25.78. How much money does he have now? _____

20. Shakira went to the supermarket. She bought eggs for $1.54, milk for $2.28, and brownie mix for $3.29. How much did Shakira pay altogether? _____

Place Value and Decimals

LESSON 3.2

Look at the number in the table, and answer the following questions.

hundreds	tens	ones	.	tenths	hundredths	thousandths
3	6	2	.	5	8	4

1. What does the 3 stand for? _____
2. What does the 6 stand for? _____
3. What does the 2 stand for? _____
4. What does the 5 stand for? _____
5. What does the 8 stand for? _____
6. What does the 4 stand for? _____

Write <, >, or = to complete each statement.

7. 0.8 ◯ 0.08
8. 1.01 ◯ 1.10
9. 6 ◯ 6.6
10. 0.79 ◯ 7.9
11. 0.5 ◯ 0.50
12. 1.25 ◯ 0.125

Order the following numbers from least to greatest.

13. 5.32, 5.23, 5.4 _____
14. 0.084, 0.09, 0.0356 _____
15. 4.13, 4.03, 4.3 _____
16. 8.9, 8.74, 8.09 _____

Comparing and Ordering Decimals

Fill in the ◯ with <, >, or =.

1. 8.9 ◯ 8.8
2. 308 ◯ 3.08
3. 7.15 ◯ 7.51
4. 57.4 ◯ 57.8
5. 407 ◯ 4.70
6. 2.0 ◯ 20
7. 3.50 ◯ 3.5
8. 43.43 ◯ 34.34
9. 102.02 ◯ 102.2
10. 70.0 ◯ 70
11. 6,000 ◯ 600.0
12. 36.08 ◯ 38.06
13. 8.0087 ◯ 8.008
14. 52.798 ◯ 52.788
15. 389.987 ◯ 398.879

Write these numbers in order from least to greatest.

16. 3.89, 3.7, 2.98 _____
17. 56.92, 56.29, 52.96 _____
18. 2, 0.2, 1.2 _____
19. 77.008, 77.8, 87.09 _____
20. 2.897, 2.89, 2.98 _____
21. 101.11, 110.1, 11.001 _____
22. 3.050, 3.005, 3.6 _____
23. 0.009, 0.0089, 0.01 _____

Solve these problems.

24. Joseph, Ariel, and DaShaun ran the fifty-yard dash together. Joseph's time was 8.325 seconds. Ariel's time was 9.4 seconds. DaShaun's time was 8.235 seconds.

    a. Who had the fastest time? _____

    b. Put the race times in order from least to greatest. _____

25. Kaitlyn had $19.75, and Sajev had $37.50. Kaitlyn said that she had more money than Sajev. Is she correct? Explain.

_____

24 Chapter 3 • *Decimals*

Real Math • Grade 5 • *Practice*

LESSON 3.4

Name _____ Date _____

Adding and Subtracting Decimals

Add or subtract. Watch the signs.

①   9.42
   + 6.92

②   10.67
   − 3.84

③   67.80
   − 1.304

④   92
   − 43.07

⑤   84.6
   + 97.329

⑥   32.768
   + 3.376

⑦   84.5
   + 6.5

⑧   83.40
   − 16.75

⑨   293.6
   − 87.4

⑩   103.92
   + 87.4

⑪   167.82
   − 92.4

⑫   792.86
   − 92.85

⑬   0.009
   + 0.006

⑭   0.416
   + 0.819

⑮   0.942
   + 7.697

⑯   9.875
   − 0.324

⑰   32.7
   − 8.004

⑱   134.3
   − 98.7

⑲   0.924
   − 0.039

⑳   262.0
   − 84.3

Solve these problems.

㉑ Rachel earned $18.36 on her first day at work. On her second day she earned $24.50. How much more did she earn the second day?

_____

㉒ Mr. Clark has $6.50 in his wallet. The lunch he wants to order costs $7.95. Can he buy lunch? Explain.

_____

Real Math • Grade 5 • *Practice*   Chapter 3 • *Decimals*

LESSON 3.5

Name _____ Date _____

Applying Math

Solve these problems.

1. Rinku bought two books and one magazine at the bookstore. She gave the clerk $20.00 and received $5.48 in change. What was the total cost of Rinku's purchases? _____

2. Nora has a collection of 129 arrowheads. She is buying some boxes for the arrowheads. Each box can hold 9 arrowheads. How many boxes does she need? _____

3. Some time ago, Zachary started doing stomach crunches every day. He began by doing 10 crunches per day, and he is now doing 50 crunches each day. For how long has Zachary been exercising? _____

4. David has $275 in his savings account. If he withdraws $135 for a new bicycle, how much money will be left in his account? _____

5. Kristen earns $3,000 per month. How much money does she earn in one year? _____

6. Caleb bought a pineapple for $2.19, a bag of cherries for $3.39, and a bunch of bananas for $0.81. How much did the fruit cost altogether? _____

7. Mr. Alvarez had 50 bicycle flags. He said that he gave 19 to students in the fourth grade, 21 to students in the fifth grade, and 26 to students in the sixth grade. Was that possible? _____

8. The Super Fresh Market has 5-pound boxes of Shiny Bright laundry detergent on sale for $3.95. The regular price of the same box of Shiny Bright is $3.59 at the Royal Beagle Store. Which store has the better buy? _____

9. Lara walks 14 blocks from home to school each day. Blocks in her town are about 200 meters long. About how many meters does she walk from home to school? _____

10. Chelsey is making drapes for the 4 windows in her living room. She needs 4 meters of fabric for the drapes for each window. How much fabric does Chelsey need altogether? _____

LESSON 3.6

Multiplying and Dividing Decimals

Find each product.

1. 713 × 10 = _____
2. 7.13 × 10 = _____
3. 71.3 × 100 = _____
4. 2.552 × 10 = _____
5. 2.552 × 100 = _____

6. 2.552 × 1,000 = _____
7. 0.99 × 10 = _____
8. 0.099 × 10 = _____
9. 0.99 × 100 = _____
10. 0.099 × 1,000 = _____

Find each quotient.

11. 782 ÷ 10 = _____
12. 7.82 ÷ 10 = _____
13. 78.2 ÷ 100 = _____
14. 0.874 ÷ 10 = _____
15. 874 ÷ 1,000 = _____

16. 0.097 ÷ 10 = _____
17. 0.097 ÷ 100 = _____
18. 23.05 ÷ 10 = _____
19. 23.05 ÷ 100 = _____
20. 23.05 ÷ 1,000 = _____

Multiply or divide. Watch the signs.

21. 602 × 10 = _____
22. 232 × 100 = _____
23. 136 ÷ 1,000 = _____
24. 91.4 ÷ 100 = _____
25. 20.2 ÷ 100 = _____
26. 7.54 × 10 = _____
27. 8.18 × 1,000 = _____
28. 87.5 × 10 = _____

29. 201 ÷ 100 = _____
30. 0.577 × 1,000 = _____
31. 11.3 ÷ 1,000 = _____
32. 5.23 ÷ 10 = _____
33. 3.02 × 100 = _____
34. 0.85 × 1,000 = _____
35. 0.511 ÷ 100 = _____
36. 192 ÷ 10,000 = _____

Lesson 3.7

Metric Units

Name _____ Date _____

> **Remember:** The prefix *milli-* means "one-thousandth of."
> The prefix *centi-* means "one-hundredth of."
> The prefix *deci-* means "one-tenth of."
> The prefix *kilo-* means "one thousand."

Use the above information to answer the following.

1. How many milligrams are in 1 gram? _____
2. How many centigrams are in 1 gram? _____
3. How many decimeters are in 1 meter? _____
4. How many meters are in 1 kilometer? _____
5. Which is bigger, a kilogram or a milligram? _____

Find the missing measure.

6. 26 m = _____ cm
7. 104 mm = _____ cm
8. 54 dm = _____ m
9. 1.9 kg = _____ g
10. 4,700 g = _____ kg
11. 3 L = _____ mL
12. 3.5 m = _____ cm
13. 800 mL = _____ L
14. 0.6 L = _____ mL
15. 59 g = _____ kg

16. 600 mL = _____ L
17. 39 kg = _____ g
18. 16 g = _____ kg
19. 210 mg = _____ g
20. 51 mm = _____ cm
21. 84 mm = _____ m
22. 800 cm = _____ m
23. 64 mL = _____ L
24. 700 mm = _____ m
25. 49 cm = _____ m

LESSON 3.8

Name _____ Date _____

Choosing Appropriate Metric Measures

Choose the unit of measure that makes the most sense.

| millimeter (mm) | kilometer (km) | centimeter (cm) |
| milliliter (mL) | meter (m) | liter (L) |

1. length of a pencil _____
2. height of a building _____
3. distance between cities _____
4. amount of water in a bathtub _____
5. amount of soda in a can _____
6. width of your finger _____
7. amount of liquid in a test tube _____
8. distance from home plate to center field _____

Ring the measurement that makes the most sense.

9. The bottle of juice holds (2 mL   2 L).
10. A dime is about (1 mm   1 cm   1 km) thick.
11. A spoon holds (5 mL   5 L) of liquid.
12. A sheet of paper is (3 cm   30 cm   300 cm) long.
13. Lindsay walks (6 km   60 km   600 km) along a nature trail.
14. A can holds (50 mL   350 mL   850 mL) of liquid.
15. A football field is about (10 m   100 m   1,000 m) long.

Solve this problem.

16. James has an apple that weighs 10 grams. Alicia has a bag of grapes that weighs 1 pound. Whose fruit weighs more? _____

Real Math • Grade 5 • Practice   Chapter 3 • Decimals **29**

LESSON 3.9

Multiplying Decimals by Whole Numbers

Multiply. Check your answers to see if they make sense.

① 2.25
 × 13

② 225
 × 1.3

③ 22.5
 × 13

④ 8.08
 × 47

⑤ 80.8
 × 47

⑥ 0.436
 × 22

⑦ 436
 × 2.2

⑧ 4.36
 × 22

⑨ 2.008
 × 111

⑩ 2008
 × 1.11

⑪ 7.13
 × 92

⑫ 0.713
 × 92

⑬ 71.3
 × 92

⑭ 32.6
 × 49

⑮ 326
 × 4.9

Solve these problems.

⑯ Last week, 250 bags of pretzels were sold at the cafeteria. Each bag cost $0.65. How much was spent on pretzels last week? _____

⑰ Emily earns $35.75 a week for babysitting. If she works for 12 weeks, how much money will she earn? _____

⑱ Troy grilled 15 pieces of chicken for his family's cookout last weekend. If each piece of chicken cost $1.25, how much did Troy spend altogether? _____

LESSON 3.10

Name _____   Date _____

Rounding and Approximating Numbers

Round each number to the nearest ten.

1. 38 _____
2. 335 _____
3. 54 _____
4. 49 _____
5. 902 _____
6. 264 _____
7. 8,260 _____
8. 1,005 _____
9. 537 _____

Round each number to the nearest hundred.

10. 184 _____
11. 608 _____
12. 899 _____
13. 2,113 _____
14. 5,476 _____
15. 3,550 _____
16. 4,735 _____
17. 2,796 _____
18. 9,385 _____

Round each number to the nearest thousand.

19. 35,255 _____
20. 9,478 _____
21. 9,803 _____
22. 72,874 _____
23. 60,844 _____
24. 39,194 _____
25. 73,525 _____
26. 66,549 _____
27. 14,359 _____

Round each number to the nearest whole number.

28. 78.4 _____
29. 2.60 _____
30. 4.56 _____
31. 26.1 _____
32. 5.06 _____
33. 4.3 _____
34. 31.9 _____
35. 4.5 _____
36. 7.28 _____

Solve the following problem without using a calculator or paper and pencil.

37. Angie wants to hang shelves that are 80 centimeters wide along a basement wall that measures 7 meters long. About how many shelves will Angie be able to hang side-by-side on this wall? _____

Real Math • Grade 5 • *Practice*   Chapter 3 • *Decimals* **31**

LESSON 3.11

Approximation Applications

Solve these problems without finding exact answers.

Circle City has four schools. There are 295 students at Small Edge School, 897 students at Big Rock School, 891 students at Round Top School, and 898 students at Square Side School. All the schools are open five days a week, from Monday to Friday.

1. About how many students are in Circle City altogether? _____

2. Hot lunches are served to all the students at Big Rock School every day. Each student gets a carton of milk with lunch. About how many cartons of milk are served with lunches in one week? _____

3. There are 31 classrooms at Round Top School. About how many desks must be in each classroom if there are about the same number of students in each classroom? _____

4. Small Edge School serves breakfast 3 days per week to all of its students. Each breakfast comes with a banana or an orange. About how many pieces of fruit does the school serve at breakfast each week? _____

5. About 100 of the students at Small Edge School walk to school.

    a. About how many students at Small Edge School don't walk to school? _____

    b. Of those students who don't walk, how many take the bus? _____

6. Square Side School has 38 classrooms. Each classroom has 32 desks.

    a. About how many desks does Square Side School have? _____

    b. About how many desks are not being used? _____

    c. If about 350 students enter Square Side School next year and only about 200 students leave, will there be enough desks? _____

32 Chapter 3 • *Decimals*   Real Math • Grade 5 • *Practice*

LESSON 3.12

Name _____ Date _____

Understanding Decimal Division Problems

Divide. Do not use remainders. All the exercises have exact decimal answers.

1. 5)12
2. 5)9
3. 5)2.5
4. 5)2.8

5. 4)1
6. 8.1 ÷ 9 = _____
7. 45 ÷ 6 = _____
8. 8)60

9. 9.3 ÷ 2 = _____
10. 3)9.3
11. 6)12.72
12. 8)64.24

13. 2 ÷ 4 = _____
14. 4)2.4
15. 3.9 ÷ 2 = _____
16. 3.9 ÷ 5 = _____

17. 3)1.23
18. 1.26 ÷ 6 = _____
19. 2)7
20. 17 ÷ 2 = _____

Solve these problems.

21. Miguel has 23 baseball cards and wants to split them with 4 of his friends. If each person gets an equal amount of cards, how many cards will each person get? How many cards will be left over? _____

22. Mrs. Scott bought 46 stickers to give to students in her 2 classes. If there are 24 students in each class, will she have enough stickers for everyone? Explain. _____

23. Miara wants to buy a skirt for $25.75 and a shirt for $16.98. About how much money does she need? _____

Real Math • Grade 5 • Practice   Chapter 3 • Decimals 33

LESSON 3.13

Name _____ Date _____

Interpreting Quotients and Remainders

Solve the following problems by using division.

1. Andrea paid $4.95 for 9 packages of flower seeds. If each package was the same price, how much did 1 package cost? _____

2. Mr. Tillman is ordering canned fruit for the school cafeteria. He needs to buy at least 156 cans. The cans of fruit are only sold in cases of 8. How many cases should he order? _____

3. Rachel needs 75 hot dogs for a Fourth of July picnic. There are 10 hot dogs in a package. How many packages should she buy? _____

4. Mackenzie has 20 yards of rope. If she cuts it into 8 equal pieces, how long will each piece be? _____

5. Lupé bought 2 notebooks apiece for herself and 3 friends when she went shopping. Altogether, the notebooks cost $10.32. How much money does each of Lupé's friends owe her? _____

6. JaShawn and Audra want to make cookies for the school picnic. The recipe for a dozen cookies calls for 2.5 cups of flour. How many cups of flour would be needed to make 6 dozen cookies? _____

7. There are a dozen boxes of nails in each carton. Each box of nails weighs 1.15 kilograms. How much would 1 carton of nails weigh? _____

8. Steven has a road-racing track of a total length of 64.5 inches. If each piece of straight track measures 5.375 ($5\frac{3}{8}$) inches, how many pieces of straight track laid end-to-end does Steven have? _____

9. Jennifer prepared 12 quarts of punch for a holiday party. Assuming that each person drinks about 0.25 quart of punch, how many people can Jennifer serve? _____

10. Soonkyoung has 8 boards, each of which is 3.5 meters long. She needs several strips of wood 2 meters long for a cabinet she is building. How many pieces of wood can Soonkyoung cut from the boards she has? _____

LESSON 3.14

Name _____ Date _____

Decimals and Multiples of 10

Divide. Give exact answers. Check your answers to see if they make sense.

① 5)45   ② 50)450   ③ 500)4,500   ④ 50)4,500

⑤ 50)45   ⑥ 8)56   ⑦ 80)560   ⑧ 80)56

⑨ 40)36   ⑩ 400)3,600   ⑪ 5)65   ⑫ 50)65

⑬ 50)6,500   ⑭ 50)6,550   ⑮ 60)969   ⑯ 80)640

⑰ 80)6.4   ⑱ 5)44   ⑲ 20)35   ⑳ 40)300

Solve. Watch the signs. Use shortcuts when you can.

㉑ 45.9 ÷ 1,000 = _____

㉒ 45.9 × 100 = _____

㉓ 45.9 ÷ 10 = _____

㉔ 937 ÷ 10 = _____

㉕ 93.7 ÷ 1,000 = _____

㉖ 1.25 × 10 = _____

㉗ 1.25 ÷ 10 = _____

㉘ 1.25 ÷ 100 = _____

㉙ 0.351 × 10,000 = _____

㉚ 3.51 × 100 = _____

Solve these problems.

㉛ Mr. Williamson needs to buy 1,800 pencils. Pencils are sold in boxes of 30 each. How many boxes does he need to order? _____

㉜ Mariana needs 10 pounds of watermelon and 10 pounds of cantaloupe to make fruit salad. Watermelon costs $1.09 per pound, and cantaloupe costs $0.98 per pound. How much money does Mariana need altogether? _____

Real Math • Grade 5 • *Practice*   Chapter 3 • *Decimals* **35**

LESSON 3.15

Name _____ Date _____

Applying Decimals

Solve these problems.

1 Renika can buy engines for her model rockets at the Hobby Shop for $0.35 each. At the Model Mart, she can buy a package of 10 engines for $3.15. She needs to buy 20 engines.

   a. At which store will she get a better buy? _____

   b. How much money will she save? _____

2 Abby wants to knit 24 dish towels for her friend's new apartment. Each towel requires 0.85 yards of material. How much material should she buy? _____

3 Keith's golf club is 1.43 meters long. Julia's golf club is 134 centimeters long.

   a. Whose golf club is longer? _____

   b. How much longer is it? _____

4 Dena bought 2 tablets of paper for $1.85 each and 5 pencils for 18¢ each. How much did the paper and pencils cost altogether? _____

5 Mrs. Morgan needs to buy fencing for her vegetable garden. The garden is in the shape of a rectangle measuring 12.2 meters long by 9.1 meters wide.

   a. How much fencing does Mrs. Morgan need to buy? _____

   b. Mrs. Morgan has set aside a budget of $150 to buy the fencing. What is the most she can spend per meter of fencing? _____

6 Admission to the space and industry museum is $1.75 for children and $2.50 for adults. Can a group of 5 adults and 7 children get into the museum for under $25? _____

7 How much would it cost to buy 10 posters for $2.69 each? _____

36 Chapter 3 • *Decimals*   Real Math • Grade 5 • *Practice*

LESSON 4.1

Name _____ Date _____

Using Your Calculator

See how fast you can find the answers by timing yourself. Solve without a calculator first; then use a calculator to find the answers. Are these problems easier with or without a calculator?

Time without calculator: _____   Time with calculator: _____

1. 10 × 47 = _____
2. 100 × 47 = _____
3. 1,000 × 47 = _____
4. 10,000 × 47 = _____
5. 100,000 × 47 = _____
6. 10 + 47 = _____
7. 1,000 + 47 = _____
8. 32,000 ÷ 10 = _____

9. 320 ÷ 10 = _____
10. 610,000 ÷ 1,000 = _____
11. 6,100,000 ÷ 10,000 = _____
12. 4,250 × 0 = _____
13. 4,250 + 0 = _____
14. 7 × 5 = _____
15. 36 ÷ 9 = _____
16. 46 − 16 = _____

Solve the following exercises three times, using each of these rules.

A. Perform the operations in order from left to right.

B. Do multiplication and division first; then do any addition and subtraction.

C. Do addition and subtraction first; then do any multiplication and divison.

17. 7 × 8 + 2 = _____
18. 12 − 6 ÷ 3 = _____
19. 5 + 9 − 4 = _____
20. 32 ÷ 4 − 3 = _____

21. 2 + 7 × 3 = _____
22. 3 × 4 × 8 = _____
23. 24 ÷ 6 + 2 = _____
24. 30 − 10 ÷ 5 = _____

25. Does the order in which you perform the operations matter? Explain.

_____
_____

Real Math • Grade 5 • *Practice*   Chapter 4 • *Function Rules* 37

Lesson 4.2 — Using Number Patterns to Predict

Read the table below. Ring the numbers you will hit.

	If you start at this number:	And you keep doing this:	Will you hit these numbers?		
1	0	Add 6	a. 18	b. 50	c. 72
2	3	Add 2	a. 15	b. 50	c. 65
3	6	Add 4	a. 14	b. 30	c. 64
4	80	Subtract 5	a. 55	b. 28	c. 11
5	200	Subtract 9	a. 182	b. 164	c. 119
6	1,000	Subtract 150	a. 800	b. 700	c. 500
7	10	Add 3	a. 21	b. 35	c. 55
8	1	Add 5	a. 9	b. 24	c. 51
9	75	Subtract 3	a. 65	b. 55	c. 45
10	500	Subtract 30	a. 410	b. 330	c. 200
11	44	Add 21	a. 65	b. 86	c. 105
12	350	Subtract 16	a. 333	b. 318	c. 300
13	4,000	Subtract 375	a. 3,625	b. 3,250	c. 2,900
14	222	Add 102	a. 322	b. 426	c. 530
15	17	Add 17	a. 35	b. 50	c. 67

Chapter 4 • Function Rules

Real Math • Grade 5 • Practice

Name _____ **Date** _____

Repeated Operations: Savings Plans

Complete the table, and then answer the questions.

Maribel earns $7.00 each week. Every week she puts $3.75 into her savings account and spends the remaining $3.25. Maribel made a table to keep a record of her money.

Week	Amount Earned	Amount to Spend	Amount to Save	Amount in Savings Account
1	$7.00	$3.25	$3.75	$3.75
2	$7.00	$3.25	$3.75	$7.50
3	$7.00	$3.25	$3.75	
4	$7.00	$3.25	$3.75	
5	$7.00	$3.25	$3.75	
6	$7.00	$3.25	$3.75	
7	$7.00	$3.25	$3.75	
8	$7.00	$3.25	$3.75	
9	$7.00	$3.25	$3.75	
10	$7.00			
11	$7.00			
12	$7.00			

❶ How much will Maribel have saved at the end of 24 weeks? _____

❷ How much will Maribel have saved at the end of 1 year (52 weeks)? _____

❸ How many weeks will it take Maribel to save enough to buy a new bike that costs $199.50? _____

Real Math • Grade 5 • *Practice*   Chapter 4 • *Function Rules* **39**

Lesson 4.4 Function Machines

Complete the table for each function machine.

1) $x \xrightarrow{+6} y$

In	Out
4	10
1	
3	
12	
50	
100	
	131
	136

2) $x \xrightarrow{+30} y$

In	Out
10	
15	
40	
82	
	145
	68
103	
	1,000

3) $x \xrightarrow{+18} y$

In	Out
	18
8	
	30
	72
853	
	47
100	
	1,018

Find the number that went in (*x*), the number that came out (*y*), or an addition rule that the function machine could be using.

4) 6 →(?)→ 15 _____

5) 6 →(+14)→ *y* _____

6) *x* →(+3)→ 12 _____

7) 8 →(+28)→ *y* _____

8) 37 →(+5)→ *y* _____

9) 5 →(?)→ 23 _____

10) 1 →(+16)→ *y* _____

11) *x* →(+24)→ 54 _____

Solve.

12) A function machine gives 30 as the answer when 5 is put in. There are two ways you could program your calculator to do this. List the answers. _____

LESSON 4.5

Name _____ Date _____

Multiplication Function Rules

Find the multiplication rule in each case. Mr. Palermo has made the calculator a multiplication function machine.

Mr. Palermo pushed	The display showed	The rule is
① 5 =	15	_____
② 9 =	54	_____
③ 3 =	24	_____
④ 1 5 =	360	_____
⑤ 2 4 3 8 =	4,876	_____
⑥ 2 1 =	1,050	_____

Complete each table.

⑦ x —(×4)→ y

In	Out
2	
8	
6	
3	

⑧ x —(×12)→ y

In	Out
3	
	120
	60
6	

⑨ x —(×7)→ y

In	Out
	7
8	
	77
9	

Solve.

⑩ Ashley and her friends bought T-shirts to decorate. Ashley bought 3 T-shirts for $18, Jenny bought 5 T-shirts for $30, and Mia bought 2 T-shirts for $12.
  a. What is the cost per T-shirt? _____
  b. Mrs. Porter bought 14 T-shirts for her art class. How much did she spend? _____

Real Math • Grade 5 • *Practice*

LESSON 4.6

Name _____   Date _____

Finding Function Rules

Find a function rule that works for both pairs of numbers in each problem.

1) 10 → (?) → 40
20 → (?) → 50
? = _____

2) 5 → (?) → 50
15 → (?) → 60
? = _____

3) 5 → (?) → 30
10 → (?) → 60
? = _____

4) 0 → (?) → 0
9 → (?) → 63
? = _____

5) 4 → (?) → 20
8 → (?) → 40
? = _____

6) 0 → (?) → 0
29 → (?) → 29
? = _____

Imagine that someone has made a calculator into a function machine, but you don't know whether the calculator is using an addition rule or a multiplication rule. Each problem shows two keys that you might push and the number that would come out on the display. In each problem find the addition or multiplication rule that the calculator is using.

7) [3][=] → 9
[7][=] → 21

The rule is _____.

8) [3][=] → 22
[4][=] → 23

The rule is _____.

9) [1][=] → 9
[0][=] → 0

The rule is _____.

10) [1][=] → 12
[0][=] → 11

The rule is _____.

11) For which of Exercises 7–10 could you have determined the function rule from only the first pair of numbers given? Why?

_____

_____

LESSON 4.7

Name _____ Date _____

Subtraction Rules and Negative Numbers

Find the subtraction rule in each case. Ms. Williamson has made the calculator a subtraction function machine.

Ms. Williamson pushed	The display showed	The rule is
❶ 2 5 =	19	_____
❷ 5 0 =	21	_____
❸ 7 =	5	_____
❹ 3 1 =	22	_____

Complete the following table.

	Temperature Before Change	Temperature Change	Temperature After Change
❺	12°C	up 4°C	
❻	−8°C (8° below 0°C)	down 2°C	
❼	3°C	down 5°C	
❽	−1°C (1° below 0°C)	up 6°C	
❾	5°C	down 5°C	
❿	7°C	down 10°C	

Add or subtract. Do not use a calculator. Watch for negative numbers.

⓫ 40 + 10 = _____

⓬ 1 + 4 = _____

⓭ 40 − 10 = _____

⓮ 1 − 4 = _____

⓯ 15 + 25 = _____

⓰ 5 − 12 = _____

⓱ (−15) + 25 = _____

⓲ (−15) − 25 = _____

⓳ (−8) − 9 = _____

⓴ 35 − 35 = _____

㉑ 0 − 8 = _____

㉒ (−10) − 10 = _____

Real Math • Grade 5 • *Practice*   Chapter 4 • *Function Rules* 43

LESSON 4.8

Name _____ Date _____

Adding and Subtracting Integers

Complete each exercise.

1. $6 - 9 =$ _____
2. $(-3) + 9 =$ _____
3. $(-3) - 9 =$ _____
4. $(-3) - (-9) =$ _____
5. $(-3) + (-9) =$ _____
6. $(-9) - (-3) =$ _____
7. $(-5) - 4 =$ _____
8. $(-5) + 4 =$ _____
9. $5 - 4 =$ _____
10. $(-4) + 5 =$ _____
11. $(-3) + (-7) =$ _____
12. $(-3) - (-7) =$ _____
13. $3 + 7 =$ _____
14. $3 - 7 =$ _____
15. $(-6) - (-6) =$ _____
16. $|-2| =$ _____
17. $|-8| =$ _____
18. $|-2| - |-8| =$ _____
19. $|2| - |-8| =$ _____
20. $|-2| + |-8| =$ _____
21. $|-2| + |8| =$ _____
22. $0 - (-1) =$ _____
23. $(-25) - (-25) =$ _____
24. $(-25) + (-25) =$ _____

Answer these questions.

25. At 7:00 A.M. the temperature was −2°C. By 3:00 P.M. the temperature had risen 6°C. By 10:00 P.M. the temperature had dropped 8°C. What was the temperature at 10:00 P.M.? _____

26. Andrew has $25 in his bank account. If he withdraws $15, deposits $6, and then withdraws $30, what will be the balance in his account? _____

27. Colin has a pond in his backyard. He placed a stick in the pond and marked the height of the water. For 8 weeks, he measured in centimeters the change in the height of the water. If the level decreased, he recorded a negative number. If the level increased, he recorded a positive number. Here are his results: −1, 0, +2, +3, −1, −3, +2, −1. At the end of the 8 weeks, was the water level higher or lower than the starting level? Explain.

_____

_____

Chapter 4 • *Function Rules*

Real Math • Grade 5 • *Practice*

LESSON 4.9
Multiplying and Dividing Integers

Complete these exercises. Watch the signs.

1. $3 \times (-5) =$ _____
2. $-3 \times 5 =$ _____
3. $-3 \times (-5) =$ _____
4. $12 \div (-4) =$ _____
5. $-12 \div (-4) =$ _____
6. $-12 \div 4 =$ _____
7. $-7 + 6 =$ _____
8. $7 + (-6) =$ _____
9. $-15 + (-6) =$ _____
10. $-15 - 6 =$ _____

11. $8 \times (-8) =$ _____
12. $(-8) \times (-8) =$ _____
13. $-25 \div 5 =$ _____
14. $25 \div (-5) =$ _____
15. $(-25) \div (-5) =$ _____
16. $4 \times 8 =$ _____
17. $-8 \times (-4) =$ _____
18. $72 \div (-9) =$ _____
19. $-24 \div (-3) =$ _____
20. $-11 \times (-6) =$ _____

21. From 5:00 P.M. until midnight, the temperature decreased at a rate of $-2°C$ per hour. If the temperature at 5:00 P.M. was $5°C$, what was the temperature at midnight? _____

22. For her birthday, Keisha received a $25 gift card for a local bookstore. Keisha selected 3 books that were priced $5 each and a journal that was priced $6. Does she have enough left on the gift card to buy a magazine that costs $3.75? Explain.

_____

_____

Real Math • Grade 5 • *Practice*   Chapter 4 • *Function Rules*

Lesson 4.10 — Patterns

Look at the numbers and images below. Describe each pattern in your own words.

1) 18   13   8   3   −2   −7   −12

_____

2) 0   2   2   4   6   10   16   26   42   68

_____

3) ↕ → ↕ ← ↕ → ↕

_____

Write the missing shapes or numbers according to a pattern for each sequence. Then tell what the pattern is. Use a calculator if you wish, and round to the nearest hundredth.

4) 1   −3   9   _____   81   −243   _____   _____

_____

5) −4   −3   −1   3   11   27   _____   _____   _____

_____

6) △   _____   ⬠   _____   ⬡   _____

_____

7) 168   84   _____   21   10.5   _____   _____

_____

Describe the pattern shown below. Describe the next shape in the pattern.

8)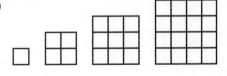

_____

46 Chapter 4 • *Function Rules*   Real Math • Grade 5 • *Practice*

LESSON 5.1 — Coordinates

Answer the following questions using the graph below.

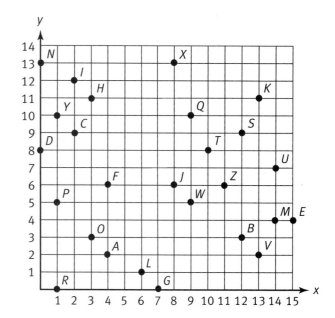

1. What are the coordinates of point *U*? _____

2. What are the coordinates of point *Q*? _____

3. What are the coordinates of point *V*? _____

4. Which three points have the same *y*-coordinate? Give the coordinates of those points. _____

5. Which four points lie on the same diagonal line? Give the coordinates of those points. _____

6. Which three points have the same *x*-coordinate? Give the coordinates of those points. _____

Write the correct letter for each of the coordinates to find the answer.

7. What city is the capital of Turkey? _____

   (4, 2); (0, 13); (13, 11); (4, 2); (1, 0); (4, 2)

8. What city is the capital of Mongolia? _____

   (14, 7); (6, 1); (4, 2); (4, 2); (0, 13); (12, 3); (4, 2); (4, 2); (10, 8); (4, 2); (1, 0)

9. What city is the capital of Indonesia? _____

   (8, 6); (4, 2); (13, 11); (4, 2); (1, 0); (10, 8); (4, 2)

Lesson 5.2 — Functions and Ordered Pairs

Complete the tables.

1 x → −8 → y

x	y
12	
18	
25	
	21
28	
	48

2 x → ×4 → y

x	y
3	
	28
6	
	36
5	
	32

3 x → ÷7 → y

x	y
7	
	3
14	
	4
35	
	6

Find three ordered pairs for each function rule, and graph them.

4 x → ÷3 → y

5 x → ×3 → y

6 x → −4 → y

48 Chapter 5 • *Graphing Functions*

Lesson 5.3 Composite Functions

Complete the table for each composite function. Then graph the ordered pairs.

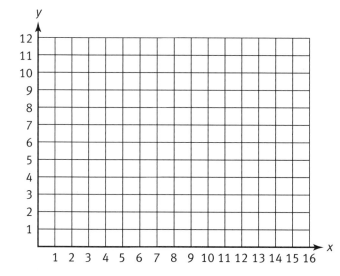

❶ x → (×3) → n → (+1) → y

x	0	0.5	1	1.5	2	3
y						

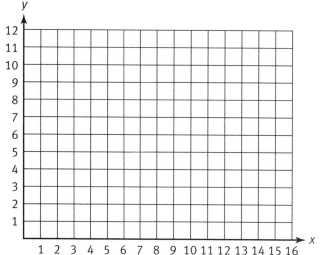

❷ x → (÷2) → n → (+2) → y

x	1	2	3	4	5	2.4
y						

Use the graphs and tables above to answer the following questions.

❸ For each exercise above, describe the figure(s) you would get if you connected the ordered pairs that you graphed.

_____

❹ Describe the first operations of the function rules in Exercises 1–2.

_____

❺ Describe the second operations of the function rules in Exercises 1–2.

_____

Real Math • Grade 5 • *Practice*   Chapter 5 • *Graphing Functions* 49

LESSON 5.4 Graphing in Four Quadrants

Complete each function machine table.

1 x —→(−6)—→ y

x	−6	−4	−2	0	2	4	6	8	10	12
y										

2 x —→(+7)—→ y

x	−12	−10	−7	−6	−4	−3	−1	0	2	3
y										

Graph each set of ordered pairs from above. Draw and label a straight line connecting the ordered pairs for each function rule.

3

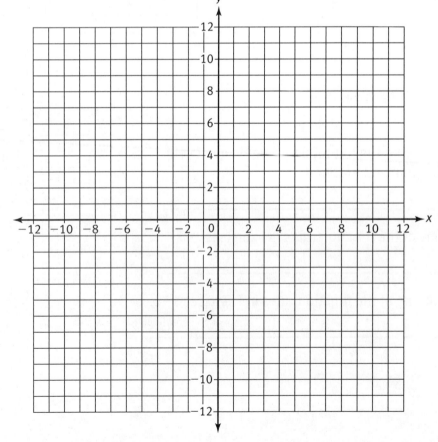

50 Chapter 5 • *Graphing Functions*

Real Math • Grade 5 • *Practice*

Name _____ **Date** _____

Making and Using Graphs

Complete the table after reading the following information.

1 Ms. Kendall owns a cookie factory. She estimates that it costs $200 each week to produce the cookies. She charges $4 per dozen cookies. To calculate her profit, Ms. Kendall uses this function rule.

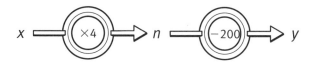

Number of dozens of cookies sold	0	5	10	20	25	40	60	75	100
Profits for the week (dollars)									

Graph the ordered pairs from the table above, and answer the following questions.

2 How many dozens of cookies must Ms. Kendall sell in order to break even for the week?

_____

3 How much profit will Ms. Kendall make if she sells 70 dozen cookies?

_____

4 If Ms. Kendall charges $5 per dozen cookies, how many dozens of cookies must she sell in order to break even for the week?

_____

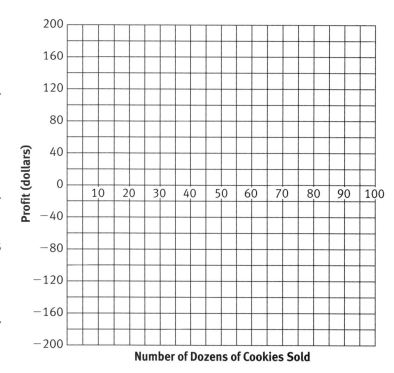

LESSON 5.6

Inverse Functions

Write the inverse operation.

① −6 _____ ② +5 _____ ③ ×21 _____

Find the value of *x* in each case. If it helps you, use the inverse function.

④ x →(−7)→ 8 ⑤ x →(×4)→ 32 ⑥ x →(+7)→ 8

_____   _____   _____

Answer each question. Keep in mind that tickets to the circus cost $6 apiece.

⑦ There are 4 people in Joe's family. How much would it cost for all of them to go to the circus?

_____

⑧ Kristin's family spent $48 for tickets to the circus. How many people are in Kristin's family?

_____

⑨ Mrs. Alvarez paid $60 for tickets for her daughter and friends to go to the circus. How many tickets did Mrs. Alvarez buy?

_____

⑩ Groups can make advance reservations for circus tickets by calling the ticket office. There is a $5 service charge added to the total cost of the tickets. Kelko's family made advance reservations and paid $41 for their tickets.

  a. What was the cost of the tickets before the service charge?

  _____

  b. How many people are in Kelko's family?

  _____

52 Chapter 5 • *Graphing Functions*   Real Math • Grade 5 • *Practice*

Lesson 5.7

Inverse of a Composite Function

Write the inverse function of each of these functions.

1. $x \xrightarrow{+4} n \xrightarrow{\div 5} y$ _____

2. $x \xrightarrow{\times 6} n \xrightarrow{-3} y$ _____

Rewrite each composite function rule as a one-step function with the same output if possible. Give the inverse of each function.

3. $x \xrightarrow{+6} n \xrightarrow{-3} y$ _____

4. $x \xrightarrow{\div 2} n \xrightarrow{\times 4} y$ _____

5. $x \xrightarrow{\times 2} n \xrightarrow{+4} y$ _____

6. $x \xrightarrow{\times 7} n \xrightarrow{-5} y$ _____

For each function, find a number that you can put in and get out the same number.

7. $\xrightarrow{\times 3} \xrightarrow{-6}$ _____

8. $\xleftarrow{-5} \xleftarrow{\times 2}$ _____

9. $\xrightarrow{\times 4} \xrightarrow{-12}$ _____

10. $\xleftarrow{-12} \xleftarrow{\times 5}$ _____

Name _____ **Date** _____

Lesson 5.8 Using Composite Functions

Answer the following questions.

Yi opened an earring shop where she makes and sells her own earrings. She charges $5.00 per pair and adds $1.50 service charge per order. Yi wrote this composite function to help her figure out how much to charge a customer.

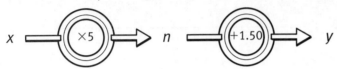

1. Yi received an order for 3 pairs of earrings. How much will this customer be charged? _____

2. A customer received a bill for $26.50. How many pairs of earrings were ordered? _____

3. On Tuesday, Yi received 3 separate orders—one for 2 pairs of earrings, one for 3 pairs of earrings, and one for 1 pair of earrings. What will be the total receipts for the day? _____

4. Lara and Lynette each wanted to buy 2 pairs of earrings. By combining their orders, how much was their bill? _____

The Drive Safely Taxi Company charges $3.00 per mile and adds a $1.05 service charge per ride. The company wrote this composite function to help the drivers figure out how much to charge a customer.

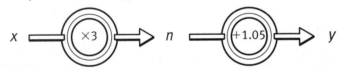

5. Mr. Ortega wants to go from his house to the airport 8 miles away. How much will the driver charge him? _____

6. Sarah pays $97.05 for a taxi ride. How many miles did she travel? _____

7. Hassan and Jillian share a ride to work, which is a distance of 5 miles away. By sharing the taxi, how much do they save altogether? _____

8. One driver received a total of $77.10 from two separate customers. How many miles did the taxi driver drive? _____

9. The taxi company's competitor charges $2.00 per ride and adds $.75 per mile. Write a composite function to help the drivers figure out how much to charge a customer.

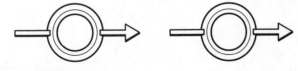

54 Chapter 5 • *Graphing Functions*   **Real Math** • Grade 5 • *Practice*

Temperature Conversions

Lesson 5.9

Answer each question. Then use the function rules to complete the table.

1. What is the function rule for converting Celsius temperatures to Fahrenheit temperatures?

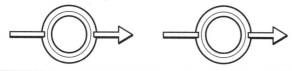

2. What is the function rule for converting Fahrenheit temperatures to Celsius temperatures?

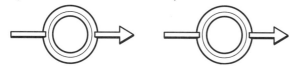

3.
Celsius temp.	−20		0	5		25			
Fahrenheit temp.		23		59			95	167	212

Find the estimating rules to complete the following table.

4. What is the estimating rule for converting Celsius temperatures to Fahrenheit temperatures?

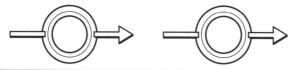

5. What is the estimating rule for converting Fahrenheit temperatures to Celsius temperatures?

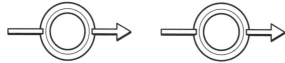

6.
Celsius temp.	−20		0	5		25			
Fahrenheit temp.		20		60			100	180	230

Standard Notation for Functions

LESSON 5.10

Complete each table.

1 $y = x + 5$

x	y
0	5
2	
5	
−1	
−4	

2 $y = x - 8$

x	y
14	6
12	
9	
5	
2	

3 $y = x + 12$

x	y
1	13
6	
12	
25	
40	

4 $y = 6x$

x	y
2	12
4	
6	
8	
11	

5 $y = \frac{x}{4}$

x	y
4	1
8	
16	
28	
36	

6 $y = \frac{x}{7}$

x	y
0	
14	
42	
70	
	11

Write a function rule for each situation.

7 Jin gets paid $8 for each hour of babysitting.

_____

8 In the Bishop family, the 3 brothers split equally the money they earned mowing lawns.

_____

9 In football, a field goal increases a team's score by 3.

_____

10 For each wrong answer on a test, 5 points out of 100 were deducted.

_____

LESSON 5.11

Composite Functions in Standard Notation

Complete each table.

1 $y = 2x + 5$

x	0	4	9	3	6
y					

4 $y = 3x - 8$

x	8	4	1	3	0
y					

2 $y = \frac{x}{3} - 3$

x	18	12	6	9	3
y					

5 $y = \frac{x}{5} + 1$

x	0	10	20	35	50
y					

3 $y = 5x - 7$

x	2	5	1	3	0
y					

6 $y = \frac{x}{2} + 8$

x	4	10	6	0	2
y					

Write a function rule for each situation.

7 A bicycle rental fee is $10 plus $2 per hour.

_____

8 Matthew opens a bank account with a $75 deposit. He plans to save $15 per month.

_____

9 A car that cost $15,000 depreciates $300 each month.

_____

Linear Equations

LESSON 5.12

Solve for *x* or *y* in each equation.

1. $4x - 32 = 4$ _____
2. $8(9) + 6x = 78$ _____
3. $2x + 5 = 23$ _____
4. $4(5) - 26 = y$ _____
5. $-11x + 70 = 4$ _____
6. $\frac{x}{7} + 9 = 5$ _____
7. $\frac{240}{6} - 25 = y$ _____
8. $5(-4) + 4 = y$ _____
9. $\frac{x}{4} \times 6 = 24$ _____
10. $-17.1 + 3(3.15) = y$ _____
11. $-11.3(7) + 29.1 = y$ _____
12. $3.5 \times 2.9 + 8.1 = y$ _____
13. $6x + 5.9 = 68.9$ _____
14. $34x - 102 = 34$ _____
15. $98(-3) + 98(3) = y$ _____
16. $13x + 7 = -58$ _____
17. $32.50 + 19.95x = 172.15$ _____
18. $-1.05x + 13.1 = 10.58$ _____
19. $-108 + 14x = -10$ _____
20. $\frac{x}{2.5} \times 17.1 = 23.94$ _____
21. $-7(2.35) - 25 = y$ _____
22. $4x + 2 = 21$ _____
23. $177 + 39x = 996$ _____
24. $987 - 45(30) = y$ _____

58 Chapter 5 • *Graphing Functions*   **Real Math** • Grade 5 • *Practice*

LESSON 6.1

Name _____ Date _____

Fractions of a Whole

Solve the following problems.

1 Mr. Ferraro divided his farm into 12 sections of equal area. He planted 1 section of tomatoes, 2 sections of lettuce, and 9 sections of cucumbers.

   a. What fraction of the farm has tomatoes planted? _____

   b. What fraction of the farm has cucumbers planted? _____

   c. What fraction of the farm has lettuce or cucumbers planted? _____

2 Meagan bought 10 sets of buttons. There were 4 sets of red, 5 sets of blue, and 1 set of green.

   a. What fraction of the sets of buttons was red? _____

   b. What fraction of the sets of buttons was green? _____

   c. What fraction of the sets of buttons was blue or green? _____

Solve for n.

3 $\frac{1}{5}$ of $40 = n$ _____

4 $\frac{2}{5}$ of $40 = n$ _____

5 $\frac{4}{5}$ of $40 = n$ _____

6 $\frac{1}{3}$ of $18 = n$ _____

7 $n = \frac{5}{5}$ of 30 _____

8 $\frac{3}{5}$ of $40 = n$ _____

9 $n = \frac{3}{4}$ of 28 _____

10 $n = \frac{1}{4}$ of 28 _____

Ring each correct answer. You may wish to draw a picture of each answer choice to help you decide the correct fraction.

11 Nadia wrote 24 short poems. There were 10 poems about animals. What fraction of Nadia's poems were about animals?

   a. $\frac{10}{12}$ c. $\frac{5}{6}$

   b. $\frac{7}{12}$ d. $\frac{5}{12}$

12 Of the 30 children on a trip, 18 brought their lunch from home. What fraction of the children on the trip brought their lunch from home?

   a. $\frac{1}{3}$ c. $\frac{3}{5}$

   b. $\frac{1}{2}$ d. $\frac{2}{3}$

Real Math • Grade 5 • *Practice*

LESSON 6.2

Fractions of Fractions

Multiply.

1. $\frac{1}{5} \times \frac{1}{3} = $ _____

2. $\frac{2}{7} \times \frac{2}{5} = $ _____

3. $\frac{6}{6} \times \frac{1}{3} = $ _____

4. $\frac{6}{7} \times \frac{3}{7} = $ _____

5. $\frac{7}{8} \times \frac{7}{8} = $ _____

6. $\frac{2}{9} \times \frac{3}{3} = $ _____

7. $\frac{4}{7} \times \frac{4}{9} = $ _____

8. $\frac{5}{7} \times \frac{1}{9} = $ _____

9. $\frac{2}{5} \times \frac{2}{3} = $ _____

10. $\frac{1}{3} \times \frac{3}{4} = $ _____

11. $\frac{2}{5} \times \frac{5}{6} = $ _____

12. $\frac{5}{8} \times \frac{1}{3} = $ _____

13. $\frac{3}{4} \times \frac{5}{7} = $ _____

14. $\frac{1}{3} \times \frac{2}{3} = $ _____

15. $\frac{7}{8} \times \frac{2}{3} = $ _____

16. $\frac{4}{5} \times \frac{1}{2} = $ _____

17. $\frac{3}{5} \times \frac{7}{7} = $ _____

18. $\frac{6}{7} \times \frac{3}{7} = $ _____

Write the correct fraction to make each statement true.

19. $\frac{1}{3}$ of $\square = \frac{4}{15}$ _____

20. $\frac{2}{5} \times \square = \frac{4}{25}$ _____

21. $\frac{3}{5} \times \square = \frac{3}{35}$ _____

22. $\frac{5}{9}$ of $\square = \frac{25}{81}$ _____

23. $\frac{7}{9} \times \square = \frac{7}{27}$ _____

24. $\frac{1}{6}$ of $\square = \frac{6}{42}$ _____

25. $\frac{2}{5}$ of $\square = \frac{4}{45}$ _____

26. $\frac{1}{8} \times \square = \frac{7}{64}$ _____

LESSON 6.3

Name _____ Date _____

Decimal Equivalents of Fractions

Give the decimal equivalent or approximation to three decimal places (to the nearest thousandth) for each exercise.

1. $\frac{6}{9}$ = _____
2. $\frac{2}{8}$ = _____
3. $\frac{5}{8}$ = _____
4. $\frac{2}{4}$ = _____
5. $\frac{7}{8}$ = _____
6. $\frac{1}{8}$ = _____
7. $\frac{1}{5}$ = _____
8. $\frac{1}{7}$ = _____
9. $\frac{4}{7}$ = _____
10. $\frac{1}{4}$ = _____
11. $\frac{2}{10}$ = _____
12. $\frac{9}{10}$ = _____
13. $\frac{5}{7}$ = _____
14. $\frac{7}{10}$ = _____
15. $\frac{3}{8}$ = _____
16. $\frac{3}{4}$ = _____
17. $\frac{5}{9}$ = _____
18. $\frac{4}{5}$ = _____
19. $\frac{8}{9}$ = _____
20. $\frac{1}{9}$ = _____
21. $\frac{5}{6}$ = _____
22. $\frac{1}{3}$ = _____
23. $\frac{4}{9}$ = _____
24. $\frac{1}{6}$ = _____

Answer each question.

25. Jessica made a cake in the shape of a rectangle. She cut it into 8 equal pieces. Jessica ate 2 pieces, her brother ate 3 pieces, and her mother ate 1 piece.

    a. What fraction of the cake was left in the pan? _____

    b. What fraction of the cake was eaten? _____

    c. Was there more cake eaten or more left in the pan? _____

Real Math • Grade 5 • *Practice*   Chapter 6 • *Fractions*

LESSON 6.4

Equivalent Fractions

Find the missing numerator or denominator.

1. $\frac{1}{4} = \frac{6}{\square}$ _____
2. $\frac{2}{5} = \frac{\square}{35}$ _____
3. $\frac{\square}{5} = \frac{15}{25}$ _____
4. $\frac{3}{4} = \frac{21}{\square}$ _____
5. $\frac{\square}{30} = \frac{5}{6}$ _____
6. $\frac{13}{39} = \frac{\square}{3}$ _____
7. $\frac{4}{7} = \frac{16}{\square}$ _____
8. $\frac{\square}{5} = \frac{12}{15}$ _____
9. $\frac{24}{32} = \frac{3}{\square}$ _____
10. $\frac{3}{8} = \frac{27}{\square}$ _____
11. $\frac{5}{8} = \frac{\square}{24}$ _____
12. $\frac{5}{\square} = \frac{40}{48}$ _____
13. $\frac{4}{9} = \frac{20}{\square}$ _____
14. $\frac{6}{\square} = \frac{1}{5}$ _____
15. $\frac{2}{\square} = \frac{22}{33}$ _____
16. $\frac{6}{\square} = \frac{30}{35}$ _____
17. $\frac{\square}{48} = \frac{3}{4}$ _____
18. $\frac{2}{9} = \frac{16}{\square}$ _____

Reduce each of the following fractions to lowest terms.

19. $\frac{10}{45} =$ _____
24. $\frac{30}{35} =$ _____
20. $\frac{8}{24} =$ _____
25. $\frac{6}{48} =$ _____
21. $\frac{14}{21} =$ _____
26. $\frac{25}{35} =$ _____
22. $\frac{27}{63} =$ _____
27. $\frac{8}{32} =$ _____
23. $\frac{12}{60} =$ _____
28. $\frac{56}{72} =$ _____

Write three equivalent fractions for each of the following fractions.

29. $\frac{1}{8}$ _____
31. $\frac{2}{6}$ _____
30. $\frac{5}{7}$ _____
32. $\frac{3}{5}$ _____

LESSON 6.5

Name _____ Date _____

Fractions with the Same Denominator

Add or subtract. Reduce answers to lowest terms where possible.

1. $\frac{1}{12} + \frac{6}{12} = $ _____
2. $\frac{4}{7} + \frac{1}{7} = $ _____
3. $\frac{3}{10} + \frac{4}{10} = $ _____
4. $\frac{3}{10} + \frac{5}{10} = $ _____
5. $\frac{2}{8} + \frac{5}{8} = $ _____
6. $\frac{8}{20} + \frac{7}{20} = $ _____
7. $\frac{6}{9} - \frac{3}{9} = $ _____
8. $\frac{5}{16} - \frac{3}{16} = $ _____
9. $\frac{11}{12} - \frac{2}{12} = $ _____
10. $\frac{3}{7} + \frac{3}{7} = $ _____
11. $\frac{2}{10} + \frac{2}{10} = $ _____
12. $\frac{2}{8} + \frac{1}{8} = $ _____
13. $\frac{2}{5} + \frac{1}{5} = $ _____
14. $\frac{1}{6} + \frac{2}{6} = $ _____
15. $\frac{7}{16} + \frac{5}{16} = $ _____
16. $\frac{1}{12} + \frac{7}{12} = $ _____
17. $\frac{5}{8} - \frac{3}{8} = $ _____
18. $\frac{10}{18} - \frac{1}{18} = $ _____
19. $\frac{5}{9} - \frac{1}{9} = $ _____
20. $\frac{4}{15} + \frac{5}{15} = $ _____
21. $\frac{2}{6} + \frac{2}{6} = $ _____
22. $\frac{3}{8} + \frac{1}{8} = $ _____
23. $\frac{2}{15} + \frac{3}{15} = $ _____
24. $\frac{6}{12} - \frac{4}{12} = $ _____
25. $\frac{3}{14} + \frac{4}{14} = $ _____
26. $\frac{3}{8} + \frac{3}{8} = $ _____
27. $\frac{3}{10} + \frac{3}{10} = $ _____
28. $\frac{2}{20} + \frac{3}{20} = $ _____
29. $\frac{5}{18} - \frac{1}{18} = $ _____
30. $\frac{1}{16} + \frac{3}{16} = $ _____

Solve the following problems. Reduce answers to lowest terms where possible.

31. While Jasmine was on vacation, she bought 25 postcards. Jasmine planned to write 5 of them each day for 5 days.

    a. What fraction of the postcards did she plan to write the first day? _____

    b. She followed the plan the first day, but the second day she got ambitious and wrote $\frac{2}{5}$ of the original number of postcards. What fraction of the postcards did she have left to write at the end of the second day? _____

32. Mr. Klemmer made 11 meatballs of equal size.

    a. David ate 3 of the meatballs, and Ligia ate 2 of the meatballs. What fraction of the meatballs did they eat altogether? _____

    b. What fraction of the meatballs was left? _____

Real Math • Grade 5 • Practice   Chapter 6 • Fractions 63

LESSON 6.6
Practice with Fractions

Answer the questions. You can use the ruler below to help you.

1. What are all the fractional measures, down to sixteenths of an inch, that are alternative ways to identify the line for $\frac{1}{4}''$? _____

2. What are all the fractional measures, down to eighths of an inch, that are alternative ways to identify the line for $\frac{6}{16}''$? _____

3. What are all the fractional measures, down to sixteenths of an inch, that are alternative ways to identify the line for $\frac{6}{8}''$? _____

4. Kia drew a line that was $\frac{3}{4}''$ long. She then extended it by $\frac{1}{16}''$. What was the final length of the line? _____

5. You measure the height of a mug to be $3\frac{1}{2}''$. How many $\frac{1}{4}$ inches is that? _____

Decide which fraction in each of the following pairs is greater. Complete the pairs with a $<$, $>$, or $=$ symbol.

6. $\frac{1}{4}$ _____ $\frac{5}{16}$

7. $\frac{7}{16}$ _____ $\frac{3}{8}$

8. $\frac{5}{8}$ _____ $\frac{1}{4}$

9. $\frac{6}{16}$ _____ $\frac{3}{8}$

10. $\frac{2}{16}$ _____ $\frac{2}{8}$

11. $\frac{3}{8}$ _____ $\frac{1}{4}$

12. $\frac{4}{16}$ _____ $\frac{1}{8}$

13. $\frac{2}{4}$ _____ $\frac{8}{16}$

14. $\frac{6}{8}$ _____ $\frac{12}{16}$

15. $\frac{3}{4}$ _____ $\frac{14}{16}$

LESSON 6.7

Comparing Fractions

Find a common denominator for each pair of fractions.

1. $\frac{7}{12}, \frac{3}{4}$ _____
2. $\frac{1}{5}, \frac{2}{3}$ _____
3. $\frac{8}{9}, \frac{3}{5}$ _____
4. $\frac{1}{2}, \frac{2}{3}$ _____
5. $\frac{5}{6}, \frac{1}{8}$ _____
6. $\frac{7}{10}, \frac{7}{12}$ _____
7. $\frac{1}{4}, \frac{3}{10}$ _____
8. $\frac{3}{8}, \frac{2}{7}$ _____
9. $\frac{5}{8}, \frac{1}{4}$ _____

Find a common denominator for each group of fractions.

10. $\frac{1}{2}, \frac{4}{9}, \frac{5}{6}$ _____
11. $\frac{1}{6}, \frac{5}{9}, \frac{1}{3}$ _____
12. $\frac{2}{5}, \frac{1}{6}, \frac{3}{4}$ _____
13. $\frac{3}{8}, \frac{1}{4}, \frac{7}{16}$ _____
14. $\frac{1}{4}, \frac{2}{3}, \frac{1}{5}$ _____
15. $\frac{2}{7}, \frac{1}{4}, \frac{5}{8}$ _____
16. $\frac{1}{2}, \frac{1}{3}, \frac{1}{4}$ _____
17. $\frac{7}{10}, \frac{23}{30}, \frac{11}{15}$ _____
18. $\frac{5}{12}, \frac{7}{18}, \frac{1}{9}$ _____

Write <, >, or = to make each comparison correct.

19. $\frac{5}{6}$ ___ $\frac{4}{5}$
20. $\frac{2}{3}$ ___ $\frac{22}{36}$
21. $\frac{4}{9}$ ___ $\frac{15}{36}$
22. $\frac{1}{2}$ ___ $\frac{9}{18}$
23. $\frac{4}{7}$ ___ $\frac{3}{8}$
24. $\frac{3}{7}$ ___ $\frac{1}{2}$
25. $\frac{13}{16}$ ___ $\frac{3}{4}$
26. $\frac{18}{24}$ ___ $\frac{3}{4}$
27. $\frac{5}{12}$ ___ $\frac{6}{10}$

Order the fractions from least to greatest.

28. $\frac{5}{12}, \frac{1}{4}, \frac{3}{4}, \frac{5}{8}, \frac{2}{9}$ _____
29. $\frac{1}{8}, \frac{7}{8}, \frac{1}{2}, \frac{9}{16}, \frac{1}{16}$ _____
30. $\frac{5}{12}, \frac{1}{3}, \frac{5}{6}, \frac{11}{12}, \frac{1}{4}$ _____

Real Math • Grade 5 • *Practice*

LESSON 6.8

Name _____ Date _____

Counting Possible Outcomes

Answer these questions by using the tree diagrams shown.

❶ How many total outcomes are there?

_____

❷ How many outcomes have hot dogs as the meal choice?

_____

❸ How many outcomes have hamburger, fruit punch, and fruit salad?

_____

```
                    Apple Juice  ─── Fruit salad
                   ╱             ╲── Potato salad
          Hot Dog ─
                   ╲             ── Fruit salad
                    Fruit punch  ─── Potato salad


                    Apple Juice  ─── Fruit salad
                   ╱             ╲── Potato salad
         Hamburger ─
                   ╲             ── Fruit salad
                    Fruit punch  ─── Potato salad
```

❹ How many outcomes have hotdog as a meal and potato salad as a side dish? _____

❺ How many combinations have fruit punch as a drink? _____

❻ How many combinations have potato salad as a side dish? _____

❼ How many combinations have fried chicken as a meal choice? _____

❽ What meal combination would you choose?

_____

Use the information below to create a tree diagram.

❾ Lila wants to know how many outfits she can make with her new clothes. She bought black pants and white pants. She bought a green sweater and a red sweater. Lila also bought sneakers and boots.

66 Chapter 6 • Fractions

LESSON 6.9

Name _____ Date _____

Probability and Fractions

Use the table to answer the following questions.

POSSIBLE OUTCOMES FOR FLIPPING A COIN FOUR TIMES

H	H	H	H		T	H	H	H
H	H	H	T		T	H	H	T
H	H	T	H		T	H	T	H
H	H	T	T		T	H	T	T
H	T	H	H		T	T	H	H
H	T	H	T		T	T	H	T
H	T	T	H		T	T	T	H
H	T	T	T		T	T	T	T

1 What is the probability of flipping a coin four times and getting all heads? _____

2 What is the probability of flipping a coin four times and getting exactly three heads? _____

3 When you flip a coin once, what is the probability of heads? _____

4 When you flip a coin three times and get HHH, what is the probability that the next flip is heads? _____

5 Suppose you flip a coin four times and get TTTT. What is the probability that on the next four flips you will get TTTT? _____

6 Enrique flips a coin three times. Trey flips a coin four times. Who has a higher probability of getting all heads? _____

7 There are four possible outcomes for flipping a coin two times, eight possible outcomes for flipping a coin three times, and sixteen possible outcomes for flipping a coin four times. How many possible outcomes are there for flipping a coin five times? How can you describe the pattern?

_____

Answer the following questions about a deck of forty cards numbered 1 to 40.

8 What is the probability that the top card in the deck is 32? _____

9 What is the probability that a card you pick is a multiple of eight? _____

10 What is the probability that a card you pick is even? _____

Real Math • Grade 5 • Practice   Chapter 6 • Fractions **67**

LESSON 6.10

Name _____ Date _____

Adding Fractions

Add the fractions. Write your answers in lowest terms.

1. $\frac{3}{9} + \frac{1}{3} =$ _____
2. $\frac{1}{10} + \frac{1}{3} =$ _____
3. $\frac{3}{4} + \frac{1}{16} =$ _____
4. $\frac{7}{12} + \frac{1}{4} =$ _____
5. $\frac{3}{5} + \frac{3}{10} =$ _____
6. $\frac{3}{18} + \frac{2}{9} =$ _____
7. $\frac{1}{6} + \frac{1}{2} =$ _____
8. $\frac{2}{7} + \frac{1}{4} =$ _____
9. $\frac{7}{12} + \frac{1}{8} =$ _____
10. $\frac{1}{2} + \frac{1}{10} =$ _____
11. $\frac{3}{5} + \frac{1}{4} =$ _____
12. $\frac{6}{9} + \frac{1}{6} =$ _____
13. $\frac{2}{8} + \frac{1}{2} =$ _____
14. $\frac{7}{8} + \frac{1}{16} =$ _____
15. $\frac{4}{7} + \frac{1}{14} =$ _____
16. $\frac{1}{3} + \frac{1}{8} =$ _____
17. $\frac{1}{4} + \frac{1}{5} =$ _____
18. $\frac{4}{8} + \frac{1}{32} =$ _____

Answer the following questions.

19. Ms. Haddad used $\frac{1}{5}$ of her garden for tomatoes and $\frac{1}{4}$ for cucumbers. What fraction of her garden did Mrs. Haddad use for tomatoes and cucumbers altogether? _____

20. In a marathon, $\frac{1}{4}$ of the runners were 20 years old or younger, $\frac{4}{7}$ were between 21 and 40, and the rest were over 40. What fraction of the marathon runners were over 40 years old? _____

21. Ginny swam $\frac{1}{3}$ km. After resting, she swam an additional $\frac{3}{8}$ km. How far did Ginny swim altogether? _____

22. Carlos budgeted $\frac{1}{4}$ of his pay for savings, $\frac{1}{3}$ for clothes, and the rest for entertainment. What fraction of his budget was for entertainment? _____

LESSON 6.11

Name _____   Date _____

Subtracting Fractions

Subtract. Write your answers in lowest terms.

① $\frac{7}{9} - \frac{1}{3} =$ _____   ② $\frac{3}{5} - \frac{3}{10} =$ _____   ③ $\frac{1}{2} - \frac{2}{8} =$ _____

④ $\frac{2}{7} - \frac{1}{4} =$ _____   ⑤ $\frac{3}{4} - \frac{1}{7} =$ _____   ⑥ $\frac{5}{10} - \frac{2}{5} =$ _____

⑦ $\frac{1}{2} - \frac{1}{6} =$ _____   ⑧ $\frac{7}{8} - \frac{1}{16} =$ _____   ⑨ $\frac{1}{4} - \frac{1}{5} =$ _____

⑩ $\frac{5}{8} - \frac{1}{4} =$ _____   ⑪ $\frac{1}{3} - \frac{1}{10} =$ _____   ⑫ $\frac{2}{9} - \frac{3}{18} =$ _____

⑬ $\frac{3}{4} - \frac{1}{5} =$ _____   ⑭ $\frac{3}{5} - \frac{1}{4} =$ _____   ⑮ $\frac{9}{15} - \frac{2}{5} =$ _____

⑯ $\frac{2}{4} - \frac{1}{3} =$ _____   ⑰ $\frac{4}{9} - \frac{3}{18} =$ _____   ⑱ $\frac{4}{12} - \frac{1}{6} =$ _____

Solve these problems.

⑲ Two identical tank trucks are carrying liquid propane. The first truck is $\frac{2}{3}$ full, and the second is $\frac{1}{8}$ full. The second truck transferred its entire load to the first truck. How full is the first truck now, expressed as a fraction?   _____

⑳ A swimming pool was filled to its full capacity until it started to leak. After one month, it had lost $\frac{1}{5}$ of its water. The next month it lost $\frac{3}{10}$ of the original amount. The third month it lost $\frac{1}{6}$ of the original amount.

   a. What fraction of the original amount of water was lost by the end of the second month?   _____

   b. What fraction of the original amount of water was lost by the end of the third month?   _____

㉑ Mr. Nguyen had an important project at work that had to be completed. In one 24-hour period, he worked 14 hours and slept 7 hours.

   a. What fraction of the 24-hour period did Mr. Nguyen spend sleeping?   _____

   b. What fraction of the 24-hour period did Mr. Nguyen spend either working or sleeping?   _____

Real Math • Grade 5 • *Practice*

LESSON 6.12

Name _____ Date _____

Applying Fractions

Add or subtract.

1. $\frac{7}{8} - \frac{1}{3} =$ _____
2. $\frac{2}{3} - \frac{3}{10} =$ _____
3. $\frac{15}{16} - \frac{1}{8} =$ _____
4. $\frac{1}{9} + \frac{1}{6} =$ _____
5. $\frac{3}{4} + \frac{1}{6} =$ _____
6. $\frac{1}{5} + \frac{4}{7} =$ _____
7. $\frac{5}{7} - \frac{1}{3} =$ _____
8. $\frac{1}{6} + \frac{1}{4} =$ _____
9. $\frac{5}{10} - \frac{1}{5} =$ _____

Solve these problems.

10. Mallory sliced a pepperoni pizza into 6 equal slices and a mushroom pizza into 8 equal slices. Ramón had 1 slice of pepperoni pizza and 1 slice of mushroom pizza. What fraction of a whole pizza did Ramón eat? _____

11. Catherine had a ribbon that was $\frac{7}{12}$ yd long. She used $\frac{1}{4}$ yd to decorate a wall hanging in her bedroom. How much ribbon was left? _____

12. In a relay race, 4 runners ran equal distances over a 1-mile course.

    a. How far did each runner run? _____

    b. Suppose the first runner ran just $\frac{1}{16}$ of a mile and the other three runners ran equal distances in the 1-mile race. How far would each of the other runners run? _____

13. Sandeep finished $\frac{1}{3}$ of a jigsaw puzzle before dinner and another $\frac{1}{7}$ of the puzzle after dinner. What fraction of the jigsaw puzzle does he have left to do? _____

14. Mr. Yoshida used $\frac{1}{9}$ of a tank of gas before driving to visit his uncle. After his trip, he had $\frac{1}{2}$ of a tank of gas left. How much gas did Mr. Yoshida use to visit his uncle? _____

15. June has 30 days, and July has 31 days. Andrew worked for 20 days in June and for 20 days in July. Did Andrew work the same fraction of each month? Explain.

_____

LESSON 7.1

Mixed Numbers and Improper Fractions

Replace each mixed number with an equivalent improper fraction.

1. $1\frac{3}{4} =$ _____
2. $2\frac{3}{8} =$ _____
3. $1\frac{5}{8} =$ _____
4. $4\frac{2}{3} =$ _____
5. $1\frac{5}{7} =$ _____
6. $2\frac{3}{5} =$ _____
7. $3\frac{1}{5} =$ _____
8. $4\frac{1}{2} =$ _____
9. $4\frac{1}{4} =$ _____
10. $1\frac{1}{6} =$ _____
11. $4\frac{3}{7} =$ _____
12. $3\frac{7}{8} =$ _____
13. $6\frac{1}{9} =$ _____
14. $1\frac{2}{3} =$ _____
15. $11\frac{3}{4} =$ _____
16. $8\frac{1}{4} =$ _____
17. $7\frac{2}{5} =$ _____
18. $10\frac{5}{7} =$ _____

Replace each improper fraction with an equivalent mixed number.

19. $\frac{11}{4} =$ _____
20. $\frac{13}{6} =$ _____
21. $\frac{17}{3} =$ _____
22. $\frac{13}{8} =$ _____
23. $\frac{16}{7} =$ _____
24. $\frac{21}{5} =$ _____
25. $\frac{45}{7} =$ _____
26. $\frac{21}{6} =$ _____
27. $\frac{17}{4} =$ _____
28. $\frac{9}{5} =$ _____
29. $\frac{17}{7} =$ _____
30. $\frac{19}{8} =$ _____
31. $\frac{16}{5} =$ _____
32. $\frac{12}{7} =$ _____
33. $\frac{65}{9} =$ _____
34. $\frac{113}{10} =$ _____
35. $\frac{53}{6} =$ _____
36. $\frac{77}{8} =$ _____

LESSON 7.2

Multiplying Mixed Numbers

Multiply. Check to see that your answers make sense.

1. $2\frac{1}{3} \times \frac{1}{2} =$ _____
2. $1\frac{1}{4} \times 1\frac{1}{8} =$ _____
3. $4\frac{1}{5} \times 2\frac{1}{3} =$ _____
4. $3\frac{1}{2} \times \frac{1}{4} =$ _____
5. $4\frac{2}{3} \times \frac{1}{3} =$ _____
6. $3\frac{3}{4} \times 2\frac{2}{3} =$ _____
7. $1\frac{1}{8} \times 2\frac{1}{3} =$ _____
8. $1\frac{3}{8} \times 1\frac{1}{4} =$ _____
9. $2\frac{2}{5} \times \frac{1}{2} =$ _____
10. $1\frac{3}{5} \times \frac{1}{3} =$ _____
11. $3\frac{3}{4} \times 1\frac{1}{3} =$ _____
12. $2\frac{5}{6} \times 1\frac{1}{3} =$ _____
13. $2\frac{2}{3} \times \frac{1}{4} =$ _____
14. $\frac{3}{4} \times 1\frac{1}{4} =$ _____
15. $1\frac{1}{2} \times 2\frac{1}{3} =$ _____
16. $\frac{5}{8} \times 1\frac{1}{3} =$ _____
17. $\frac{7}{8} \times 1\frac{3}{7} =$ _____
18. $1\frac{1}{4} \times \frac{4}{9} =$ _____
19. $2\frac{1}{4} \times 1\frac{2}{3} =$ _____
20. $1\frac{1}{3} \times 1\frac{1}{2} =$ _____

Solve these problems.

21. Clarissa baby-sat $3\frac{1}{2}$ hours on Friday. She is paid $6 per hour. How much did Clarissa earn? _____

22. Mrs. Youn is making 4 matching costumes for her children. Each costume requires $2\frac{3}{4}$ yards of fabric. How many yards of fabric does Mrs. Youn need to buy? _____

23. There is a $\frac{1}{4}$ mile racetrack at Dani's school. She ran around it $9\frac{1}{2}$ times before stopping for water. How many miles did she run? _____

LESSON 7.3

Adding Mixed Numbers

Add.

1. $4\frac{2}{3} + 3\frac{1}{5} = $ _____

2. $1\frac{1}{6} + 1\frac{8}{9} = $ _____

3. $2\frac{1}{5} + 1\frac{3}{10} = $ _____

4. $2\frac{1}{6} + 5\frac{2}{3} = $ _____

5. $3\frac{2}{5} + 4\frac{1}{2} = $ _____

6. $3\frac{3}{4} + 2\frac{1}{8} = $ _____

7. $3\frac{1}{3} + 4\frac{3}{4} = $ _____

8. $3\frac{2}{3} + 1\frac{1}{2} = $ _____

9. $7\frac{1}{8} + 6\frac{3}{4} = $ _____

10. $5\frac{3}{7} + 4\frac{2}{3} = $ _____

11. $2\frac{3}{4} + 1\frac{3}{8} = $ _____

12. $1\frac{7}{20} + 6\frac{7}{10} = $ _____

13. $3\frac{4}{9} + 1\frac{1}{3} = $ _____

14. $5\frac{1}{6} + 5\frac{1}{8} = $ _____

15. $4\frac{7}{12} + 8\frac{4}{5} = $ _____

16. $7\frac{1}{6} + 3\frac{1}{7} = $ _____

17. $3\frac{7}{9} + 2\frac{2}{3} = $ _____

18. $12\frac{7}{9} + 6\frac{17}{18} = $ _____

Solve these problems.

19. Machine A can produce $5\frac{9}{20}$ parts each minute. Machine B can produce $2\frac{5}{16}$ parts each minute. How many parts can both machines produce each minute? _____

20. Stock prices are shown as fractions. A stock price on the New York Stock Exchange opened at $\$63\frac{7}{8}$. During the day, it gained $\$2\frac{3}{4}$. What was the closing price of this stock? _____

21. Jennifer bought $1\frac{1}{4}$ pounds of bean sprouts, $2\frac{1}{2}$ pounds of mushrooms, and $3\frac{3}{8}$ pounds of carrots. How many pounds of vegetables did Jennifer buy altogether? _____

22. Kareem lives $3\frac{1}{2}$ blocks from school. Bianca lives $2\frac{5}{6}$ blocks farther from school than Kareem. How far from school does Bianca live? _____

LESSON 7.4

Name _____ Date _____

Subtracting Mixed Numbers

Subtract.

1. $8\frac{7}{10} - 4\frac{2}{5} = $ _____
2. $6\frac{3}{8} - 6\frac{1}{4} = $ _____
3. $3\frac{1}{2} - 1\frac{1}{4} = $ _____
4. $5\frac{2}{3} - 1\frac{7}{9} = $ _____
5. $4\frac{1}{10} - 1\frac{1}{5} = $ _____
6. $5\frac{1}{8} - 1\frac{3}{4} = $ _____
7. $9\frac{11}{12} - 5\frac{5}{6} = $ _____
8. $10\frac{2}{3} - 4\frac{1}{9} = $ _____
9. $13\frac{7}{8} - 5\frac{2}{3} = $ _____
10. $4\frac{1}{4} - 2\frac{5}{8} = $ _____
11. $9\frac{1}{2} - 5\frac{7}{10} = $ _____
12. $11\frac{14}{15} - 4\frac{3}{10} = $ _____

Add or subtract. Watch the signs.

13. $1\frac{1}{3} + 4\frac{1}{3} = $ _____
14. $5\frac{3}{4} - 1\frac{1}{8} = $ _____
15. $1\frac{1}{3} + 4\frac{1}{3} = $ _____
16. $5\frac{5}{8} - 3\frac{3}{8} = $ _____
17. $5\frac{1}{8} - 1\frac{3}{4} = $ _____
18. $1\frac{7}{20} + 2\frac{4}{5} = $ _____
19. $3\frac{1}{6} + 1\frac{1}{2} = $ _____
20. $5\frac{6}{7} + 2\frac{2}{3} = $ _____
21. $4\frac{5}{8} - 2\frac{3}{4} = $ _____

Solve these problems.

22. A carpenter has a board in his workshop that is $5\frac{3}{8}$ inches wide, but it is $2\frac{1}{4}$ inches wider than what he needs for his project. How wide does the board need to be for the carpenter's job?

_____

23. This year, Emily spends $6\frac{3}{10}$ hours in school each day. Last year, she spent $5\frac{7}{12}$ hours in school each day. How much more time does Emily spend each day in school this year than she did last year?

_____

24. Anja had $4\frac{1}{3}$ yards of fabric. Then she cut $1\frac{4}{5}$ yards to make a sash. How much fabric does Anja have left?

_____

LESSON 7.5

Name _____ Date _____

Addition and Subtraction Applications

Solve these problems.

① The time is now half past four.

  a. What time will it be in $\frac{1}{2}$ hour? _____

  b. What time will it be in $2\frac{1}{2}$ hours? _____

  c. What time will it be in $3\frac{1}{4}$ hours? _____

② Jonathan is $12\frac{1}{2}$ years old.

  a. How many years ago was he $2\frac{1}{2}$ years old? _____

  b. How many years ago was he 9 years old? _____

  c. In how many years will he be $15\frac{1}{2}$ years old? _____

  d. In how many years will he be 21 years old? _____

③ Jasmine kept the following record of the time she spent doing homework each day last week:

Day	S	M	T	W	Th	F	S
Hours	$\frac{1}{2}$	$1\frac{1}{4}$	$1\frac{1}{2}$	$1\frac{3}{4}$	$2\frac{1}{2}$	$\frac{1}{2}$	2

  a. Altogether, how much time did Jasmine spend doing homework last week? _____

  b. How many more hours did Jasmine spend doing homework on Thursday than on Wednesday? _____

  c. Did Jasmine spend more time doing homework on Friday and Saturday, or on Tuesday and Wednesday?

  _____

④ A wire is $4\frac{7}{16}$ feet long.

  a. Suppose $\frac{9}{16}$ foot of wire is used. How much wire is left? _____

  b. Of the remaining wire in Part a, how much wire is left if an additional $2\frac{3}{16}$ feet of wire is used? _____

Real Math • Grade 5 • Practice   Chapter 7 • Mixed Numbers and Improper Fractions

LESSON 7.6

Name _____ Date _____

Dividing Fractions

Divide. Reduce when possible.

1. $15 \div \frac{3}{5} =$ _____
2. $24 \div \frac{3}{8} =$ _____
3. $9 \div \frac{6}{7} =$ _____
4. $\frac{3}{8} \div \frac{3}{4} =$ _____
5. $\frac{1}{8} \div \frac{5}{7} =$ _____
6. $6\frac{1}{5} \div 7 =$ _____
7. $\frac{5}{3} \div \frac{4}{7} =$ _____
8. $\frac{7}{12} \div \frac{1}{6} =$ _____
9. $1\frac{1}{3} \div 5 =$ _____
10. $\frac{12}{25} \div \frac{3}{5} =$ _____
11. $\frac{3}{16} \div \frac{7}{8} =$ _____
12. $3\frac{7}{8} \div \frac{5}{6} =$ _____
13. $\frac{7}{9} \div \frac{2}{3} =$ _____
14. $3\frac{1}{2} \div 2 =$ _____
15. $\frac{2}{3} \div \frac{3}{5} =$ _____

Fill in the value that will make each statement true.

16. $\frac{4}{3} \times \square = 1$ _____
17. $4 \times \square = 20$ _____
18. $30 \div \square = 5$ _____
19. $12\frac{1}{2} \div \square = 5$ _____
20. $\frac{3}{7} \div \frac{5}{8} = \square$ _____
21. $\frac{3}{8} \times \frac{3}{5} = \square$ _____
22. $\square \div \frac{5}{6} = 8$ _____
23. $\square \div \frac{3}{20} = 15$ _____
24. $\square \div 3\frac{3}{16} = 4$ _____

Solve these problems.

25. Darian measured the length of his stride as $\frac{4}{5}$ yard. At this rate, how many strides will it take him to travel 12 yards? _____

26. Elias is designing hats for a school play. Each hat requires $\frac{3}{4}$ yard of fabric. How many hats will Elias be able to make from a bolt of fabric $6\frac{1}{2}$ yards long? _____

Fractions and Decimals

LESSON 7.7

Solve the following problems in two ways—by finding a common denominator and by using decimals to approximate. For each problem, show that the two answers you get are equivalent or nearly equivalent. The first one has been done for you.

Problem	Find a Common Denominator	Change to Decimals (Seven Places)	Show the Answers are Equivalent
$\frac{1}{8} + \frac{1}{3}$	$\frac{3}{24} + \frac{8}{24} = \frac{11}{24}$	$0.125 + 0.3333333 = 0.4583333$	$\frac{11}{24} = 0.4583333$
❶ $\frac{5}{8} + \frac{1}{9}$			
❷ $\frac{3}{5} - \frac{1}{2}$			
❸ $\frac{1}{4} + \frac{1}{6}$			
❹ $\frac{2}{3} + \frac{1}{2}$			
❺ $\frac{4}{9} - \frac{1}{3}$			
❻ $\frac{7}{8} - \frac{1}{4}$			
❼ $\frac{3}{5} + \frac{1}{3}$			
❽ $\frac{5}{9} - \frac{1}{8}$			
❾ $\frac{7}{8} + \frac{3}{4}$			
❿ $\frac{5}{6} - \frac{1}{3}$			

Real Math • Grade 5 • *Practice*   Chapter 7 • *Mixed Numbers and Improper Fractions*

Decimal Equivalents of Rational Numbers

LESSON 7.8

Solve each problem in two ways. For each problem check to see that your two answers are equivalent or nearly equivalent. Round your answers to four decimal places.

1. $2\frac{1}{3} + 3\frac{1}{3} =$ _____
2. $3\frac{1}{8} + 1\frac{7}{8} =$ _____
3. $\frac{1}{2} - \frac{1}{4} =$ _____

4. $\frac{2}{3} - \frac{1}{6} =$ _____
5. $5\frac{2}{3} - 3\frac{1}{6} =$ _____
6. $\frac{7}{8} + \frac{5}{8} =$ _____

7. $3\frac{1}{2} - 2\frac{1}{3} =$ _____
8. $3\frac{2}{3} - 1\frac{7}{9} =$ _____
9. $4\frac{1}{2} + \frac{5}{8} =$ _____

10. $7\frac{1}{9} + 6\frac{1}{3} =$ _____
11. $2\frac{1}{7} - \frac{5}{7} =$ _____
12. $2\frac{1}{7} + \frac{5}{7} =$ _____

13. $6\frac{1}{4} + 6\frac{1}{3} =$ _____
14. $3\frac{2}{9} + 4\frac{2}{3} =$ _____
15. $10\frac{1}{2} - 5\frac{1}{4} =$ _____

16. $7\frac{8}{9} - 4\frac{2}{3} =$ _____
17. $3\frac{3}{4} - \frac{5}{8} =$ _____
18. $\frac{7}{8} - \frac{5}{16} =$ _____

19. $\frac{2}{3} + \frac{3}{4} =$ _____
20. $\frac{3}{7} - \frac{2}{5} =$ _____
21. $\frac{3}{4} - \frac{1}{3} =$ _____

22. $\frac{2}{3} - \frac{2}{3} =$ _____
23. $\frac{1}{8} + \frac{3}{4} =$ _____
24. $\frac{3}{7} + \frac{2}{3} =$ _____

25. $\frac{5}{6} + \frac{3}{4} =$ _____
26. $3\frac{5}{6} + 4\frac{3}{4} =$ _____
27. $5\frac{7}{12} - 1\frac{3}{4} =$ _____

28. $\frac{7}{8} + \frac{5}{8} =$ _____
29. $\frac{1}{3} - \frac{1}{9} =$ _____
30. $\frac{1}{4} - \frac{1}{16} =$ _____

LESSON 7.9

Name _____ Date _____

Using Mixed Numbers

Answer each question.

① Samuel is training for a 5-kilometer race. He wants to run an equal distance on each of 4 days this week, covering a total of 25 kilometers. If Samuel is to meet his goal, how far should he run each day that he trains? _____

② Mark's recipe for coconut cream pie calls for $\frac{2}{3}$ cup of sugar. He wants to make 4 of these pies to serve at a party. How many cups of sugar will Mark need to make them all? _____

③ Nayeli is trotting her horse Charger around a $1\frac{1}{4}$ mile track.

   a. Is it possible for Nayeli to take the horse around the track $\frac{3}{4}$ time? If so, how far will they have gone? Explain.

   _____
   _____

   b. Is it possible for Nayeli to take Charger around the track $\frac{5}{3}$ times? If so, how far will they have gone? Explain.

   _____
   _____

④ Andrew works 5 days each week. He drives $10\frac{2}{3}$ miles to get to the office and the same distance to get back home again each day.

   a. Altogether, how many miles does Andrew drive each week getting to and from work? _____

   b. Andrew's car can travel about 300 miles on a full tank of gas. About what fraction of his fuel tank's capacity does he use each week traveling to and from work? _____

⑤ Can a 2-gallon bucket that has $1\frac{1}{2}$ gallons of water in it hold an additional $\frac{11}{8}$ gallons of water? Explain.

_____

Real Math • Grade 5 • *Practice*   Chapter 7 • *Mixed Numbers and Improper Fractions*

LESSON 8.1

Averages

Find the averages.

1. 7, 8, 9, 10, 11 _____
2. 7, 9, 11, 13, 15 _____
3. 70, 72, 74, 76, 78 _____
4. 20, 21, 21, 22, 23, 25 _____
5. 40, 41, 43, 44, 47 _____
6. 8, 14, 15, 21, 22, 25 _____
7. 60, 64, 68, 68, 70 _____
8. 203, 207, 210, 211, 214 _____

Use the table to answer these questions.

JaShaun kept track of his test scores in a table like this. A perfect score is 100.

9. What was the average for all JaShaun's test scores?

10. What was JaShaun's average score in math?

11. What was JaShaun's average score in Spanish?

12. What score would JaShaun need on his next science test to have an average of 80?

Date	Test Subject	Score
4/4	Math	95
4/4	Grammar	84
4/5	Science	78
4/7	Spanish	100
4/12	Math	88
4/13	Science	79
4/14	Grammar	79
4/15	Spanish	95
4/22	Math	92
4/25	Science	65
4/25	Grammar	50
4/27	Spanish	95
4/27	Math	100
4/29	Spanish	89
4/29	Grammar	75
5/6	Math	85
5/6	Science	90
5/6	Spanish	80

LESSON 8.2

Name _____ Date _____

Mean, Median, Mode, and Range

Study the bowling scores. Then answer the questions that follow.

Bowler	Pointers	Pinsters	Alley Cats
1	75	125	88
2	91	98	130
3	127	106	187
4	133	98	116
5	127	142	93

❶ What is the range of scores for all bowlers? Explain. _____

❷ What is the mode of the Pinsters' scores? _____

❸ What is the median score for the Alley Cats? _____

❹ Which team has a mean score of about 114? _____

❺ What is the clipped range of scores for bowlers on the Pointers? How did you find it?

_____

Read each situation. Then answer the questions.

❻ Dena got the following scores on her spelling tests: 65, 65, 90, 100, 95, and 93. Which would give a more accurate picture of Dena's spelling average—mean or mode? Explain.

_____

❼ Weekend sales by size at The Shirt Shack were as follows: 137 smalls, 688 mediums, 903 larges, and 416 extra larges. The manager must reorder shirts. Which measure— mean, median, mode, or range—would best help her to decide how many of each size to order? Why?

_____

❽ In the last seven years, Yolanda has hit the following number of home runs: 17, 21, 20, 16, 24, 25, and 22. Which measure—mean, median, mode, or range—would best describe her home-run hitting performance? Explain.

_____

Real Math • Grade 5 • *Practice*

LESSON 8.3

Name _____ Date _____

Interpreting Averages

Read and think about each situation. Decide whether the mean, median, or mode would be most appropriate. Explain your answer.

1 Five players form the offensive line of the Wilmington Wolves football team. The starting five players weigh 225, 270, 215, 235, and 270 pounds. The Wolves' team booklet claims that the average starting weight of the offensive line is 270 pounds—the heaviest in the league. What measure of average allows the Wolves to make this claim? How accurate is this claim? Explain.

_____

2 Commuters complain that the M17 bus is too crowded. They want an extra bus added to the route. Here are the numbers of M17 bus riders for each day last week: Sunday: 1,046; Monday: 4,856; Tuesday: 5,138; Wednesday: 4,762; Thursday: 3,988; Friday: 4,066; Saturday: 1,825.

The city says that if the M17 records an average of 4,000 riders per day, they will add another bus to the route. Commuters claim that an average of 4,066 people rides the M17 bus each day. However, the city says that the average number of daily M17 riders is only 3,669 and that another M17 bus is not needed. How is each group finding its "average"?

_____

3 Quincy hopes to win the attendance award for his school district. In the 9 years at his school district, he has been absent 2, 0, 1, 0, 3, 0, 2, 0, and 1 days. Mira also hopes to win this award. Her mean number of days absent is 0.75 days per year over 9 years. What measure of average should Quincy use to make his best case for winning the award? Explain.

_____

4 Sasha and her family enjoy playing miniature golf. (In this game, the lowest score wins.) They are regular players at Tee Time's challenging hill course. The following are the scores for each member of the family over three rounds of play:

Sasha: 67, 71, 65
Avery: 78, 68, 77
Leslie: 66, 66, 73
Jim: 63, 79, 70

What method of finding the average would give each player his or her lowest average score? Explain.

_____

_____

82 Chapter 8 • *Division and Ratios* Real Math • Grade 5 • *Practice*

LESSON 8.4

Name _____ Date _____

Ratios and Rates

Write the best way to report the information in each statement.

1 April has thirty days, and May has thirty-one days.
_____

2 Shelly's Shoe Shop buys sandals for $8 a pair and sells them for $24 a pair.
_____

3 Jacob's family has two cars. It costs them 18¢ per mile to drive their compact car. It costs them 45¢ a mile to drive their large van.
_____

4 It took Gwen 12 minutes to do her math homework. It took Devon $\frac{1}{4}$ hour to do the same math assignment.
_____

Use a calculator to solve. Check your answers.

5 A 24-ounce bag of pretzels costs $2.89, whereas a 16-ounce bag costs $2.19.

   a. How much do the pretzels cost per ounce in the 24-ounce bag? _____

   b. How much do the pretzels cost per ounce in the 16-ounce bag? _____

   c. Which bag is the better buy? _____

6 Mr. Gutierrez drove 900 miles in 15 hours and used 28.5 gallons of gasoline.

   a. What was Mr. Gutierrez's average speed? _____

   b. At this speed, how far could Mr. Gutierrez have gone in 22 hours? _____

   c. On average, about how many miles did Mr. Gutierrez's car go for each gallon of gasoline? _____

   d. At this rate, about how many gallons of gasoline would Mr. Gutierrez need to travel another 1,200 miles? _____

Real Math • Grade 5 • *Practice*   Chapter 8 • *Division and Ratios*

LESSON 8.5
Comparing Ratios

Compare each pair of ratios. Write <, >, or =.

1. $\frac{3}{5}$ ____ $\frac{4}{5}$
2. $\frac{5}{7}$ ____ $\frac{5}{8}$
3. $\frac{5}{8}$ ____ $\frac{5}{6}$
4. $\frac{23}{43}$ ____ $\frac{24}{44}$
5. $\frac{21}{35}$ ____ $\frac{3}{5}$
6. $\frac{44}{100}$ ____ $\frac{45}{99}$
7. $\frac{4}{9}$ ____ $\frac{3}{8}$
8. $\frac{1}{4}$ ____ $\frac{2}{5}$
9. $\frac{3}{4}$ ____ $\frac{78}{100}$
10. $\frac{7}{11}$ ____ $\frac{1}{3}$
11. $\frac{23}{24}$ ____ $\frac{25}{26}$
12. $\frac{1}{2}$ ____ $\frac{189}{378}$

Write <, >, or =. Compare the division problems as you would compare ratios. You do not have to calculate the quotients.

13. 7 ÷ 4 ____ 7 ÷ 3
14. 291 ÷ 7 ____ 231 ÷ 36
15. 17 ÷ 4 ____ 17 ÷ 3
16. 1,000 ÷ 25 ____ 2,000 ÷ 50
17. 64 ÷ 4 ____ 63 ÷ 4
18. 1,000 ÷ 20 ____ 4,000 ÷ 80
19. 64 ÷ 4 ____ 65 ÷ 4
20. 900 ÷ 5 ____ 500 ÷ 9
21. 500 ÷ 6 ____ 700 ÷ 8
22. 32 ÷ 4 ____ 56 ÷ 8
23. 281 ÷ 7 ____ 291 ÷ 7
24. 200 ÷ 67 ____ 200 ÷ 76

Solve.

25. On Tuesday evening, Marina read $\frac{5}{6}$ of her social studies assignment. Jack read $\frac{4}{5}$ of his social studies assignment that evening.

    a. Which student read more of the assignment? _____

    b. If it was a 30-page assignment, how many pages did each student read?

    _____

26. So far this season, the Albion Patriots have lost 7 of 23 games played, and the Elko Lizards have lost 6 out of 21 games.

    a. Which team has the better record? _____

    b. Suppose the Patriots win their next 5 games and the Lizards win their next 4 games. How would the two records compare then?

    _____

LESSON 8.6

Name _____ Date _____

Using Approximate Quotients

Use the information to answer the problems that follow.
The table presents data about three neighboring counties.

Type of Information	Sierra	Greene	Parsons
Population	39,930	13,080	24,685
Houses/Apartments	8,579	2,571	4,072
Banks	57	43	19
Pharmacies	104	21	33
Hotels/Motels	26	71	14

① In Greene County, about how many people are there per pharmacy? _____

② About how many people are there for each bank in Sierra County? _____

③ In Parsons County, about how many times as many houses/apartments are there as hotels/motels? _____

④ In which county are there about 325 residents per bank? _____

⑤ Which county has about twice as many pharmacies as banks? _____

⑥ Which county has about 6 people per house or apartment? _____

⑦ Evergreen Bank wants to open a branch in one of these counties. Which one should they choose? Explain.

_____

_____

Write a division problem you could use to approximate the answer in each situation.

⑧ How could you seat 237 guests at 42 round tables?

_____

⑨ At a marathon 8,307 runners competed, and 213 volunteers distributed water along the course. About how many runners were there per volunteer?

_____

Real Math • Grade 5 • *Practice*   Chapter 8 • *Division and Ratios* 85

Lesson 8.7

Name _____ Date _____

Approximating Quotients

Find approximate answers.

1. 63 ÷ 31 _____
2. 146 ÷ 22 _____
3. 240 ÷ 80 _____
4. 6,444 ÷ 79 _____
5. 635 ÷ 7 _____
6. 277 ÷ 42 _____
7. 636 ÷ 7 _____
8. 242 ÷ 49 _____
9. 814 ÷ 9 _____
10. 2,463 ÷ 52 _____
11. 4,832 ÷ 62 _____
12. 1,186 ÷ 41 _____
13. 5,588 ÷ 72 _____
14. 243 ÷ 61 _____
15. 477 ÷ 79 _____
16. 4,949 ÷ 68 _____

Solve.

17. Jennifer wants to drive from Philadelphia to Phoenix, a distance of 2,081 miles. She plans to drive about 500 miles per day. If Jennifer leaves on Tuesday morning, when should she make it to Phoenix?

    _____

18. Benny's Burritos sells, on average, about 28 shrimp burritos per day. Benny has enough frozen shrimp for 215 more burritos. His next shipment arrives in 9 days. Does he have enough frozen shrimp to make it until then? Explain.

    _____

    _____

86 Chapter 8 • *Division and Ratios*

LESSON 8.8

Dividing by a Two-Digit Number

Approximate the answer to each problem, and then do the division. Round answers to the nearest whole number.

1 45)5445

2 47)3414

3 98)23456

4 62)4537

5 87)1950

6 34)7575

7 23)4892

8 87)3900

9 17)7575

Ring the correct answer. In each problem, two of the answers are clearly wrong, and one is correct.

10 26)20410
   a. 78
   b. 785
   c. 7,850

11 8)7528
   a. 941
   b. 9,411
   c. 94,110

12 325)1399775
   a. 430
   b. 437
   c. 4,307

Ring the correct answer.

13 *Population density* is a ratio that expresses the average number of people per square mile of land. In 1800 the United States had a gross area of 891,364 square miles. Its population then was 5,308,483. About how many people per square mile lived in the United States in 1800?
   a. 6   c. 600
   b. 60   d. 6,000

14 In 1900 the United States had grown in size. Its gross area was 3,618,770 square miles. Its population then was 76,212,168. What was the approximate population density in 1900?
   a. 6   c. 21
   b. 12   d. 210

Real Math • Grade 5 • *Practice*   Chapter 8 • *Division and Ratios*

LESSON 8.9

Name _____ Date _____

Practice with Division

Ring the correct answer. In each problem, two of the answers are clearly wrong, and one is correct.

1. 32)1975
   a. 617
   b. 61.7
   c. 6.17

2. 15)9150
   a. 6.1
   b. 61
   c. 610

3. 11)572
   a. 5.4
   b. 50
   c. 52

4. 7)5383
   a. 7,690
   b. 769
   c. 76.9

5. 30)9150
   a. 305
   b. 35
   c. 30.5

6. 75)975
   a. 13
   b. 19
   c. 10

7. 12)6180
   a. 5.15
   b. 51.5
   c. 515

8. 60)9150
   a. 1,525
   b. 152.5
   c. 15.26

9. 12)1560
   a. 103
   b. 165
   c. 130

Solve.

10. The elevator in an office building can safely carry a total passenger weight of 2,500 pounds. One day 16 people waited to board the elevator. What would the average weight have to be so all of them could ride safely? _____

11. Jamina buys a refrigerator. The total cost, including tax and delivery charge, is $648. She arranges to pay the store the same amount each month for 24 months. What is Jamina's monthly payment? _____

12. A pilot logged 97,617 air miles last month. During that month he was on duty for 20 days. About how many miles did the pilot fly per day? _____

13. The world's first Ferris Wheel appeared at the Chicago World's Fair of 1893. The ride had 36 wooden cars that hung from a huge steel wheel. When it was fully loaded, the Ferris Wheel could safely carry 2,160 passengers. How many people could ride in each wooden car? _____

14. Ferris Wheel operators collected about $726,800 during that World's Fair. At a cost of 50 cents per ticket, about how many tickets were sold? _____

LESSON 8.10

Name _____ Date _____

Dividing by a Three-Digit Number

Approximate first, and then divide. Round quotients to the nearest whole number.

1) 311)2488   **2)** 118)1062   **3)** 575)16100

4) 402)2814   **5)** 647)7764   **6)** 700)1234

7) 264)1355   **8)** 943)22632   **9)** 700)4277

10) 836)6886   **11)** 189)5292   **12)** 700)4877

Solve.

13) Blair earned $1,327 last winter shoveling snow. In looking back on her records, she saw that she shoveled snow on 47 days of the season. On average, how much did Blair earn for each day she shoveled snow? Round to the nearest dollar. _____

14) The indoor track at Lupe's school is an oval that is 220 yards around. (1 mile = 1,760 yards) Lupe ran 3 miles. How many times did she run around the track? _____

15) It costs $47 to buy a 12-foot bamboo tree. Mr. Miyoke wants to plant a border of bamboo trees to shade his yard. He has saved $550 for this purpose. How many bamboo trees can he buy without going over his budget? _____

16) Pioneer Stadium has 224 sections. At a recent home game, 42,355 fans packed the seats. On average, about how many fans were seated in each section? _____

Real Math • Grade 5 • *Practice*   Chapter 8 • *Division and Ratios* **89**

Batting Averages and Other Division Applications

Answer the following questions by using the table.

Thunderbirds Baseball Club Statistics			
Player	Official Times at Bat	Hits	Batting Average
Carson	59	22	
Enzo	302	64	
Jeffrey	127	39	
Kent	342	68	
Maria	84	20	
Trevor	184	61	
Vijay	246	55	

1 Find each player's batting average to the nearest thousandth, and write it in the appropriate column in the table.

2 Which player got a hit about $\frac{1}{3}$ of the times he or she was at bat? _____

3 Which player got a hit about 1 out of every 5 times at bat? _____

4 Suppose Carson had a few more hits. Would his average be closer to .400 than it is now? Explain.

_____

Solve.

5 For $456 you can get a box seat for a 12-game package. Ordinarily, box seats cost $44 per seat per game. Is it cheaper to buy a package deal or 12 individual seats? Explain.

_____

6 A 60-gram can of Fine Fin tuna costs $1.19. A 90-gram can of the same tuna costs $1.33.

    a. Which size can costs less per gram? _____

    b. About how much cheaper is it per gram? _____

Average Heights

Answer the following questions by using the table below. The table shows the heights of the ten tallest buildings in Cleveland, Dallas, and Seattle.

Building	Cleveland	Dallas	Seattle
1	947 ft	921 ft	933 ft
2	708 ft	886 ft	772 ft
3	658 ft	787 ft	740 ft
4	529 ft	738 ft	722 ft
5	450 ft	720 ft	609 ft
6	446 ft	686 ft	608 ft
7	430 ft	655 ft	605 ft
8	420 ft	645 ft	580 ft
9	419 ft	629 ft	574 ft
10	410 ft	625 ft	543 ft

❶ What is the range of heights of the ten tallest buildings in each city?

❷ To the nearest foot, what is the mean height of the ten tallest buildings in each city? _____

❸ What can you say about the mode of the data for each set of buildings in the table? Explain.

❹ Which city has the greatest median height among its ten tallest buildings?

❺ List the median heights for each city. What is the range of median heights?

❻ Choose two of the three cities. Use the data to make a double-bar graph that compares the heights of the buildings in order from tallest to shortest. Use two colors to make the bars.

LESSON 8.13

Name _____ Date _____

Using Rates to Make Predictions

Answer the following questions about continuing at the same rate.

Mr. Travis has been driving for 8 hours and has gone 497 miles.

1 At about what average speed is Mr. Travis traveling?

_____

2 About how far do you think he will have driven altogether after 12 hours?

_____

3 About how far will he have driven after 10 hours?

_____

4 About how far do you think Mr. Travis drove in the first 3 hours of his trip?

_____

5 Suppose that Mr. Travis had averaged 5 miles per hour less than he did. At that rate, about how far would he have driven after

   a. 3 hours? _____

   b. 10 hours? _____

   c. 12 hours? _____

Find the daily mean temperature for each city and complete the last column. The table shows the daily high and low temperatures recorded on March 6 in five different American cities.

	Location	Daily HIGH Temperature	Daily LOW Temperature	Daily MEAN Temperature
6	Burlington, Vermont	34°F	0°F	
7	Duluth, Minnesota	28°F	−4°F	
8	Honolulu, Hawaii	68°F	59°F	
9	Lander, Wyoming	40°F	24°F	
10	Savannah, Georgia	63°F	49°F	

92 Chapter 8 • *Division and Ratios*

Real Math • Grade 5 • *Practice*

LESSON 8.14

Name _____ Date _____

Population Density

Use a calculator and the table below to complete the following exercises.

Oneida City Elementary Schools				
School	Number of Students	Number of Computers	Number of DVD Players	Number of Library Books
Alliance	256	27	4	7,503
Hill Street	498	56	11	13,482
McAuliffe	307	25	7	10,569
Sojourner	412	41	14	9,714
Voyager	185	39	5	8,115

1 How many total computers are in Oneida City's elementary schools? _____

2 About how many students per computer are at

   a. Alliance School? _____

   b. Hill Street School? _____

   c. McAuliffe School? _____

   d. Sojourner School? _____

   e. Voyager School? _____

Answer the following questions by using a calculator and the table below.

State Population Density		
State	Population (2003)	Area in Square Miles
Idaho	1,366,332	83,570
Illinois	12,653,544	57,914
Indiana	6,195,643	36,418
Iowa	2,944,062	56,272

3 Which state has the greatest population density? _____

4 Which state has a population density of about 50 people per square mile? _____

5 What is the mean population density for these four states?

_____

Real Math • Grade 5 • Practice   Chapter 8 • Division and Ratios 93

LESSON 8.15

Name _____ Date _____

Using Ratios

Use the table to solve the problems. Round decimals to the nearest hundredth.

Endangered and Threatened Animal Species (2004)					
Animal Group	Endangered U.S.	Endangered Foreign	Threatened U.S.	Threatened Foreign	Total Species
Mammals	69	251	9	17	346
Birds	77	175	14	6	272
Reptiles	14	64	22	15	115
Amphibians	11	8	10	1	30
Fish	71	11	43	1	126
Insects	35	4	9	0	48
Crustaceans	18	0	3	0	21

1 What is the ratio of total endangered mammals to the total number of endangered and threatened mammals?

_____

2 Write a ratio that compares the number of endangered foreign amphibians to the number of all endangered amphibians.

_____

3 What is the ratio of threatened reptiles to endangered reptiles?

_____

4 Which is the greater ratio—threatened foreign birds to threatened U.S. birds or threatened foreign amphibians to threatened U.S. amphibians?

_____

5 How would you express the ratio of threatened U.S. reptiles to all threatened U.S. animal groups? Explain.

_____

_____

6 The table shows that the number of endangered and threatened insects is far less than the total number of endangered and threatened mammals. Express this relationship as a ratio.

_____

LESSON 9.1

Name _____ **Date** _____

Angles

Tell whether each angle is *acute, right, obtuse,* or *straight.* Each angle is labeled with a letter at the vertex.

①  _____   ⑤ _____

② _____   ⑥ _____

③ _____   ⑦ _____

④ _____   ⑧ _____

Complete the following exercises.

⑨ Give three different names for the angle.

_____

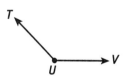

⑩ Identify and describe as many angles as possible.

_____

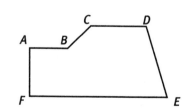

LESSON 9.2

Name _____ Date _____

Measuring Angles

Measure each angle. (Some figures may have more than one angle.) Write the name of each angle and its measure. To make it easier to read the angle, measure with your protractor. You can trace the angles and extend the sides of the angles with your ruler.

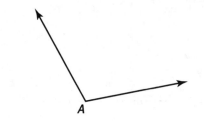

❶ Angle _____ measures _____.

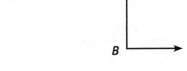

❸ Angle _____ measures _____.

❷ Angle _____ measures _____.
Angle _____ measures _____.
Angle _____ measures _____.

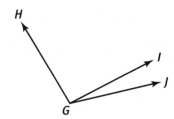

❹ Angle _____ measures _____.
Angle _____ measures _____.
Angle _____ measures _____.

❺ a. What is the measure of angle *ABC*? _____

b. What is the measure of angle *CBE*? _____

c. Explain two ways to find the measure of angle *ABE*.

_____

_____

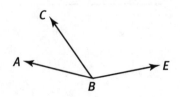

d. What is the measure of angle *ABE*? _____

96 Chapter 9 • Geometry

Real Math • Grade 5 • Practice

LESSON 9.3

Name _____ Date _____

Angles and Sides of a Triangle

Measure the three angles for each triangle. Then add the three measures to get the sum. Indicate which, if any, of the triangles are *right, isosceles,* or *equilateral*.

1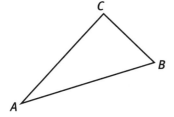

∠A = _____

∠B = _____

∠C = _____

Sum = _____

_____

3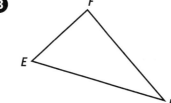

∠D = _____

∠E = _____

∠F = _____

Sum = _____

_____

2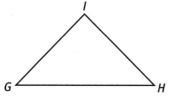

∠G = _____

∠H = _____

∠I = _____

Sum = _____

_____

4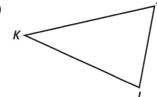

∠J = _____

∠K = _____

∠L = _____

Sum = _____

_____

Answer the question.

5 Is it possible to have a triangle with two right angles? Explain.

_____

_____

_____

Real Math • Grade 5 • *Practice*   Chapter 9 • *Geometry* **97**

LESSON 9.4

Name _____ Date _____

Drawing Triangles

Complete each exercise.

1 Draw triangle *DEF* with your ruler and protractor so that

   a. angle *F* measures 90°.

   b. side *DF* is 12 centimeters long.

   c. side *EF* is 9 centimeters long.

2 What type of triangle did you draw in Exercise 1? Explain.

_____

3 What is the length of side *DE*?

_____

4 Draw triangle *JKL* with your ruler and protractor so that

   a. the measure of angle *J* is 60°.

   b. side *JK* is 12 centimeters long.

   c. side *JL* is 12 centimeters long.

5 What type of triangle did you draw in Exercise 4? Explain.

_____

_____

6 Draw triangle *GHI* with your ruler and protractor so that

   a. the measure of angle *G* is 25°.

   b. side *GH* is 6 centimeters long.

   c. side *HI* is 2 centimeters long.

7 What is the length of side *GI*?

_____

LESSON 9.5
Congruence and Similarity

Ring the letter of the figure that is congruent to the first figure.

1   a.   b. 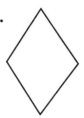 c. 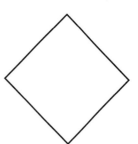 d.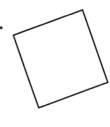

2   a.   b.   c.   d.

Record each measurement to the nearest millimeter. All of the triangles below are similar.

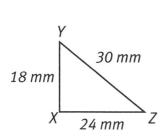
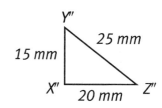

3 $z =$ _____ mm   **4** $y =$ _____ mm   **5** $x =$ _____ mm

6 $z' =$ _____ mm   **7** $y' =$ _____ mm   **8** $x' =$ _____ mm

9 $z'' =$ _____ mm   **10** $y'' =$ _____ mm **11** $x'' =$ _____ mm

Find the following ratios. Use a calculator if necessary. Round to the nearest hundredth.

12 $\dfrac{z}{y} =$ _____ mm   **13** $\dfrac{z}{x} =$ _____ mm   **14** $\dfrac{y}{x} =$ _____ mm

15 $\dfrac{z'}{y'} =$ _____ mm   **16** $\dfrac{z'}{x'} =$ _____ mm   **17** $\dfrac{y'}{x'} =$ _____ mm

18 $\dfrac{z''}{y''} =$ _____ mm **19** $\dfrac{z''}{x''} =$ _____ mm **20** $\dfrac{y''}{x''} =$ _____ mm

LESSON 9.6 Corresponding Parts of a Triangle

Write whether the triangles are similar or congruent for each pair of triangles. Also write which corresponding parts are equal. Use symbols, and make sure the letters of each triangle are in the proper order.

1.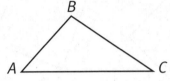

2.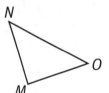

3.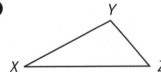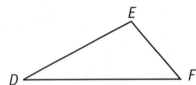

Complete the following exercises.

4. In △ABC, ∠A = 45° and ∠B = 85°. Without measuring, find the measure of ∠C.

5. △XYZ ~ △ABC. $x = 4$ cm, $y = 5$ cm, $z = 3$ cm, $b = 10$ cm, and ∠Z = 37°. Without measuring, find the length of a, the length of c, and the measure of ∠C.

6. △XTR ≅ △SVQ. $x = 4$ cm, $r = 3$ cm, and $v = 5$ cm. ∠S = 53° and ∠V = 90°. Without measuring, find the length of t, the length of s, the length of q, the measure of ∠X, and the measure of ∠Q.

Using Corresponding Parts of Triangles

Answer each question. Make sketches to help you as needed.

1 △ABC ≅ △DEF, c = 15 cm, a = 19 cm, b = 18 cm.
∠A = 65°, ∠B = 64°.

a. What is the measure of ∠C? _____   d. What is the measure of ∠E? _____

b. What is the measure of ∠F? _____   e. What is the measure of f? _____

c. What is the measure of ∠D? _____   f. What is the measure of d? _____

Find the measure of each of these angles and the length of each of these sides.

2 △ABC ~ △GHI. △ABC is the same triangle as the one in Exercise 1.
Side i = 30 cm.

a. ∠G = _____   b. ∠H = _____   c. ∠I = _____

d. i = _____   e. g = _____   f. h = _____

3 In △KLM, ∠K = 36° and ∠L = 42°. ∠M = _____

4 In △QRS, ∠Q = 45° and ∠R = ∠Q. ∠S = _____

Answer each question.

Janelle is designing a logo for her company. She is making triangles of different shapes and sizes to use in the logo. She uses a copy machine to enlarge △ABC. Janelle measures the sides of the small triangle before it is enlarged and finds a = 2 cm, b = 2 cm, and c = 3 cm.

5 Which side of the enlarged triangle do you think is the longest? _____

6 Janelle measures side b of the enlarged triangle. It is about 6 cm long. How many times shorter than that is side b of the small triangle?

_____

7 Find the lengths of side a and side c of the enlarged triangle.

a. side a of the enlarged triangle = _____

b. side c of the enlarged triangle = _____

Real Math • Grade 5 • Practice   Chapter 9 • Geometry **101**

LESSON 9.8

Name _____ Date _____

Scale Drawings

Answer each question as accurately as you can about the scale drawing of the bedroom.

1. How long is the actual dresser?
   _____

2. How long and wide is the actual bed?
   _____

3. What is the diameter of the actual wastebasket?
   _____

4. How long and wide is the top of the actual desk?
   _____

5. What is the area of the top of the actual desk?
   _____

6. How wide and deep is the actual closet?
   _____

7. What is the area of the actual room?
   _____

8. What is the area of the mirror?
   _____

9. Suppose a nightstand that measures 80 centimeters wide is brought into the room. How many squares wide will it be in the scale drawing?
   _____

10. Can the nightstand be placed in the lower left corner of the room?
    _____
    _____

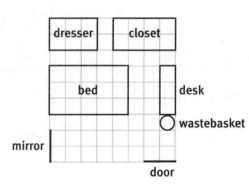

Scale: 1 ☐ = 40 cm

102 Chapter 9 • Geometry

Real Math • Grade 5 • Practice

Lesson 9.9

Using a Map Scale

Answer each question about the scale drawing of Park City shown below.

Scale: 1 cm = 0.5 km

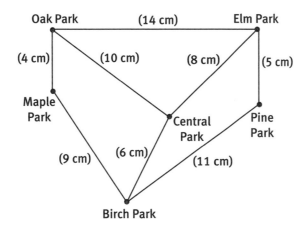

About how many kilometers is each of these actual distances in Park City?

1. Oak Park to Elm Park _____
2. Central Park to Birch Park _____
3. Pine Park to Birch Park _____
4. Maple Park to Oak Park _____
5. Maple Park to Birch Park _____
6. Elm Park to Central Park _____

On the map, how long a line segment would you draw to represent each of these distances?

7. 1.5 km _____
8. 5 km _____
9. 0.5 km _____
10. 1 km _____
11. 3.5 km _____
12. 2 km _____

13. Jerome walked in a straight line from Elm Park to Central Park and then from Central Park to Oak Park. How many kilometers did he walk altogether?

_____

14. Would Jerome have walked the same distance if he walked in a straight line from Elm Park to Oak Park without going through Central Park? Explain.

_____

Name _____   **Date** _____

Perpendicular and Parallel Lines and Quadrilaterals

Tell whether each pair of lines is *perpendicular, parallel,* or *neither.*

       ❷   ❸   ❹

_____   _____   _____   _____

Measure each of the angles of triangle *RTS*.

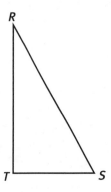

❺ ∠R = _____   ❻ ∠S = _____   ❼ ∠T = _____

Answer each question.

❽ What is the sum of the angles of triangle *RTS*? _____

❾ What is the relationship between the sum of the angles of a quadrilateral and the sum of the angles of a triangle?

_____

❿ Name at least three things in your classroom that have perpendicular lines.

_____

_____

104 Chapter 9 • Geometry   Real Math • Grade 5 • Practice

LESSON 9.11
Parallelograms

Supply the missing information. Make sketches to help you as needed.

① In quadrilateral ABCD, ∠A = 60°, ∠B = 110°, and ∠C = 100°. ∠D = _____

② In quadrilateral ABCD, ∠C = 75°, ∠D = 85°, and ∠A = ∠D. ∠B = _____

Give the standard name of each of the following figures.

③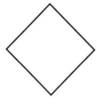
_____

④
_____

⑤
_____

⑥
_____

⑦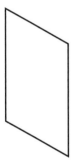
_____

⑧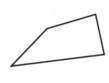
_____

⑨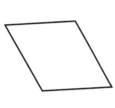
_____

⑩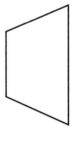
_____

Answer each question. Explain your answers.

⑪ Can an isosceles trapezoid be a parallelogram?
_____

⑫ Can a square be a rectangle?
_____

⑬ Can a square be a rhombus?
_____

⑭ Can a rhombus be a square?
_____

Real Math • Grade 5 • *Practice* Chapter 9 • *Geometry* 105

Name _____ **Date** _____

Exploring Some Properties of Polygons I

Tell whether each of these figures has a *concave* shape or a *convex* shape. If you know the names for any of the figures, write them.

1.   2.   3.

_____ _____ _____

4.   5.   6.

_____ _____ _____

Try to draw each of these figures. If it is not possible to draw the figure described, explain why.

7. a triangle with sides that measure 3, 6, and 9 centimeters
_____

8. a six-sided figure that has a concave shape
_____

9. a quadrilateral with angles that measure 110°, 80°, 90°, and 90°
_____

10. a triangle with angles that measure 40°, 70°, and 90°
_____

11. a six-sided figure that has a convex shape
_____

106 Chapter 9 • *Geometry*   **Real Math** • Grade 5 • *Practice*

LESSON 9.13

Name _____ Date _____

Exploring Some Properties of Polygons II

For each figure, determine whether it is a polygon. If it is not, explain why. If it is, identify it as *concave* or *convex*.

1

_____
_____

2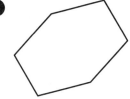

_____
_____

3

_____
_____

4

_____
_____

5

_____
_____

6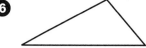

_____
_____

Answer each question.

7 What do you call a three-sided regular polygon?

_____

8 Are all squares congruent? Are they similar? Explain.

_____
_____

9 Andreas drew a regular pentagon. What is the measure of each interior angle of the pentagon?

_____

10 Charlotte drew a quadrilateral. What is the measure of each interior angle of the quadrilateral?

_____

Real Math • Grade 5 • *Practice*

LESSON 10.1

Name _____ Date _____

Circles: Finding Circumference

Given the following measures for the diameter of a circle, determine the circumference. Round to the nearest whole unit.

① 13 cm _____ ② 25 m _____ ③ 56 cm _____

④ 72 in. _____ ⑤ 2 mi _____ ⑥ 22 ft _____

Find the circumference of each circle.

⑦ 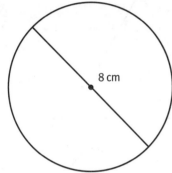 ⑧ 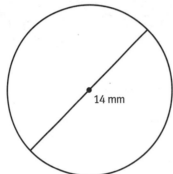 ⑨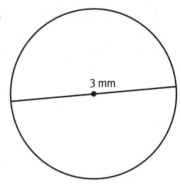

_____ _____ _____

⑩ 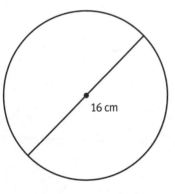 ⑪ 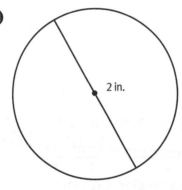 ⑫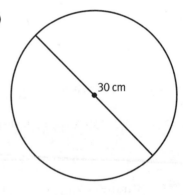

_____ _____ _____

Solve.

⑬ The circumference of a circular swimming pool is about 38 feet. What is the diameter of the pool? _____

108 Chapter 10 • *Geometry and Measurement* **Real Math** • Grade 5 • *Practice*

LESSON 10.2

Name _____ Date _____

Area of Parallelograms

Find the area of each parallelogram.

1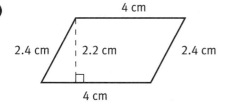

Area = _____

2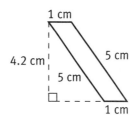

Area = _____

3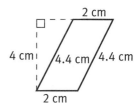

Area = _____

4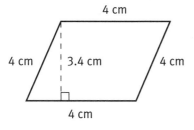

Area = _____

5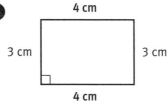

Area = _____

6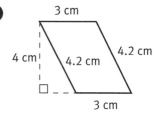

Area = _____

Answer each question.

7 How can you find the area of a parallelogram? _____

8 What additional information would you need to find the area of the parallelogram below? _____

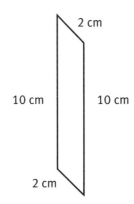

Real Math • Grade 5 • *Practice* Chapter 10 • *Geometry and Measurement*

LESSON 10.3

Name _____ Date _____

Area of Triangles

Find the area of each triangle.

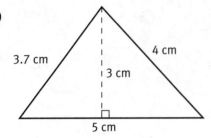

Area = _____

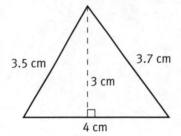

Area = _____

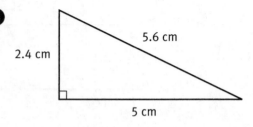

Area = _____

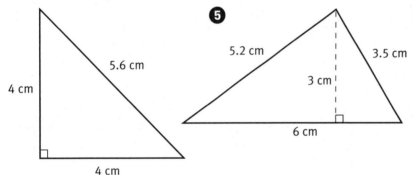

Area = _____   Area = _____

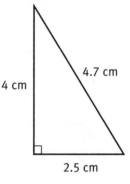

Area = _____

Find the area of the figure by using two different methods.

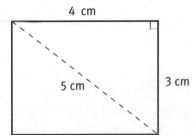

Method 1: _____

Method 2: _____

LESSON 10.4

Name _____ Date _____

Area of a Circle

Find the circumference and area of each circle. The radius is given to you. Use the formula for the area of a circle ($A = \pi r^2$), the formula for the circumference of a circle ($C = \pi d$), and an estimate of 3.14 for π to solve.

❶ $r = 500$ cm _____ ❹ $r = 8$ cm _____

❷ $r = 14$ cm _____ ❺ $r = 40$ cm _____

❸ $r = 25$ cm _____ ❻ $r = 400$ cm _____

❼ 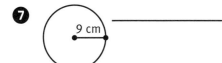 ❿

❽ 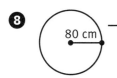 ⓫

❾ 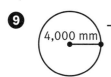 ⓬

⓭ Eric is designing a circular garden at the base of a flagpole. He wants the diameter of the garden to be one-quarter the length of the flagpole. The flagpole is 24 feet tall. What will be the circumference and area of the garden?

_____

⓮ If Eric decides to make the diameter of the garden half the length of the flagpole, what will be the circumference and area of the garden?

_____

Real Math • Grade 5 • *Practice*

LESSON 10.5

Name _____ Date _____

Area of Irregular Figures

Find the area of the figures below to the nearest hundredth of a square unit. Use π = 3.14.

1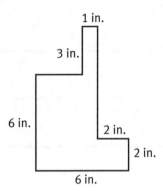

_____

2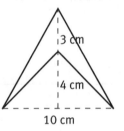

_____

3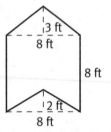

_____

4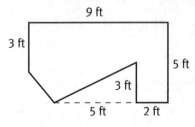

_____

5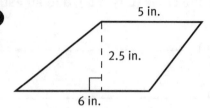

_____

6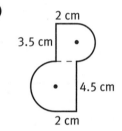

_____

7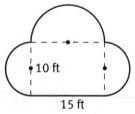

_____

8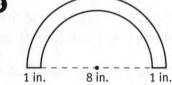

_____

112 Chapter 10 • Geometry and Measurement

Real Math • Grade 5 • Practice

LESSON 10.6

Name _____ Date _____

Rotation, Translation, and Reflection

For each of the following pairs of congruent figures, describe a combined translation and rotation that would leave one figure on top of the other.

1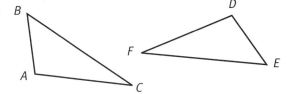

_____

_____

2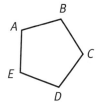

_____

_____

Trace or copy the object onto your paper, and then perform the actions described. Draw your result.

3 Reflect about line *l*. Then translate up two inches.

4 Rotate the figure 90° clockwise about point Q.

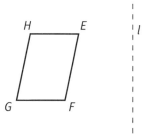

5 Describe another way to translate, rotate, or reflect the figure in Exercise 4 which yields the same result.

_____

Real Math • Grade 5 • *Practice*   Chapter 10 • *Geometry and Measurement* 113

LESSON 10.7

Name _____ **Date** _____

Symmetry

Write the number of lines of symmetry for each of the following figures, and then draw those lines of symmetry. Below each figure, write the smallest angle of rotation it has. If the angle is less than 360°, write what type of rotational symmetry the angle has.

1 _____

2  _____

3 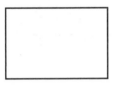 _____

4 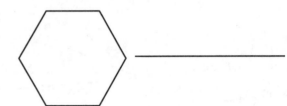 _____

5 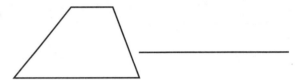 _____

6 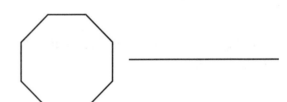 _____

Complete the table below.

Figure	Number of Sides	Number of Lines of Symmetry
equilateral triangle	_____	_____
square	_____	_____
regular pentagon	_____	_____
regular hexagon	_____	_____
regular octagon	_____	_____

7 What can you conclude about regular and irregular polygons and lines of symmetry?

_____

_____

8 How many lines of symmetry does a regular dodecagon have? Explain.

_____

_____

Name _____ **Date** _____

Paper Folding

Complete the following folding exercises.

1. Make a triangle strip that looks like the one shown below. It should have twenty equilateral triangles.

2. Fold your triangle strip to create the following shape.

3. Describe how you created the shape in the space provided. Draw a picture showing each step.

Real Math • Grade 5 • *Practice* Chapter 10 • *Geometry and Measurement* **115**

Name _____ Date _____

Making a Flexagon

Look at the design on each flexagon, and then draw all possible lines of symmetry. Write how many lines of symmetry there are altogether; then decide whether each design has rotational symmetry. If it does, describe what kind of rotational symmetry it is.

1

_____

_____

5

_____

_____

2

_____

_____

6

_____

_____

3

_____

_____

7

_____

_____

4

_____

_____

8

_____

_____

LESSON 10.10
Space Figures

Use the figures below to answer the following questions.

   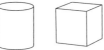

❶ Which figures have a circle as a base? _____

❷ Which figures have a polygon as a base? _____

❸ Which figures have two bases? _____

❹ Which figures have one base? _____

Trace each net. Cut it out along the outside edges. Fold along any dashed lines. Then answer the following questions about each figure you create.

a. What figure has been made by the net?
b. How many faces does the figure have?
c. How many edges does the figure have?
d. How many vertices does the figure have?

❺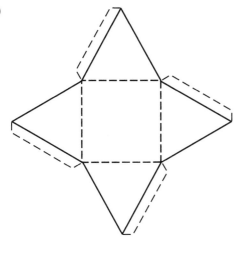

_____

❻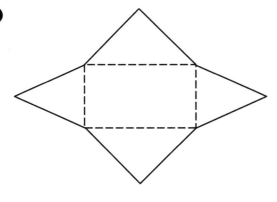

_____

❼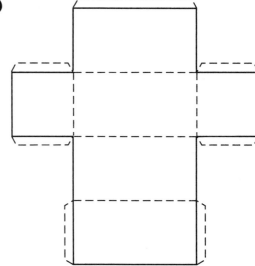

_____

❽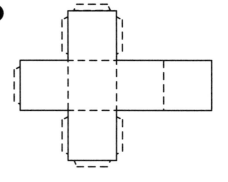

_____

Real Math • Grade 5 • *Practice*   Chapter 10 • *Geometry and Measurement* **117**

LESSON 10.11

Name _____ Date _____

Building Deltahedra

Look at the eight convex deltahedra shown below, and complete the table.

1 Use the models you've made to find out how many vertices, faces, and edges each one has; then write the numbers into the table below, where V stands for the number of vertices, E the number of edges, and F the number of faces. The first one has been done for you.

2 Fill in the rest of the table with entries for other deltahedra you and your classmates have made. If you don't know the name of a certain deltahedron, name it after the student who made it.

Name of polyhedron	V	E	F
tetrahedron	4	6	4
triangular dipyramid			
octahedron			
pentagonal dipyramid			
dodecadeltahedron			
tetracaidecadeltahedron			
heccaidecadeltahedron			
icosahedron			

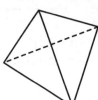
tetrahedron

triangular dipyramid

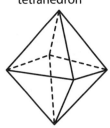
octahedron

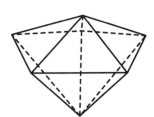
pentagonal dipyramid

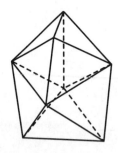
dodecadeltahedron

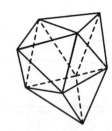
tetracaidecadeltahedron

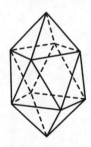
heccaidecadeltahedron

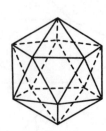
icosahedron

LESSON 10.12

Name _____ Date _____

Surface Area

Find the surface area of the boxes made from these nets. Remember to write the units in your answer.

① 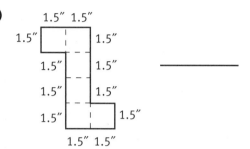 _____

② 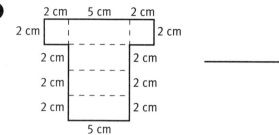 _____

Decide whether each of the following nets would make a closed box if you cut it out and folded it along the dotted line segments. If so, give the total surface area of the box. If not, write *no*.

③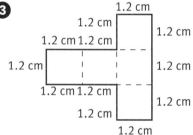

④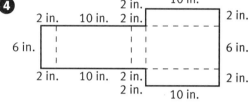

⑤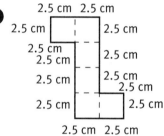

_____   _____   _____

Draw a net for each of the following figures. Calculate the total surface area for each space figure.

⑥ A cube 4 cm on each side _____

⑦ A rectangular box that is 2" by 3" by 4" _____

⑧ A rectangular box that is 0.5 cm by 1 cm by 1.5 cm _____

Answer the following questions. You might want to put cubes together to make space figures. Calculating the total length of all the edges may seem easier from the actual figure than from the net.

⑨ If the edges of a cube are 5 units long, what is the total length of all the edges? What is the total surface area of the cube? What is the volume of the cube? _____

⑩ Dara is making 5 blocks for her baby brother. She is covering the blocks with blue felt. The blocks are in the shape of a cube and measure 8 cm on each side. How much felt will she need to buy? _____

Real Math • Grade 5 • *Practice*   Chapter 10 • *Geometry and Measurement* **119**

LESSON 10.13

Volume

Name _____ Date _____

Find the volume of each solid.

This cube has a volume of 1 cubic centimeter:

①
_____

②
_____

③
_____

Find the volume in cubic units of each rectangular box.

④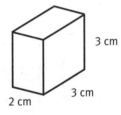
_____

⑤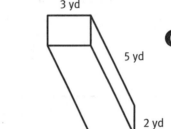
_____

⑥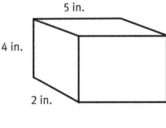
_____

Solve these problems. Be sure to use correct units.

⑦ Emma bought a new aquarium tank for her fish. The tank is 30 inches long, 12 inches wide, and 20 inches high. How many cubic inches of water will the tank hold? _____

⑧ Emma wants to leave 2 inches of space at the top of the aquarium tank. How much water will she put in the tank? _____

LESSON 11.1
Approximating Products of Decimals

Rewrite each answer, and put the decimal point in the correct place. These problems were done on a broken calculator.

1. 1.83 × 3.6 = 6588 _____
2. 1.83 × 0.36 = 6588 _____
3. 18.3 × 3.6 = 6588 _____
4. 88.6 × 2.1 = 18606 _____
5. 8.86 × 0.21 = 18606 _____
6. 88.6 × 0.21 = 18606 _____
7. 90.9 × 0.15 = 13635 _____
8. 90.9 × 1.5 = 13635 _____
9. 9.09 × 0.15 = 13635 _____
10. 0.679 × 0.154 = 104566 _____
11. 3.53 × 3.2 = 11296 _____
12. 67.9 × 15.4 = 104566 _____

Write <, >, or = to complete each exercise.

13. 2.6 × 0.8 ___ 3 × 1
14. 4.95 × 6.75 ___ 5 × 7
15. 4.0 × 5.0 ___ 4 × 5
16. 4 × 6 ___ 3.5 × 5.5
17. 0.25 × 11.85 ___ 1 × 12
18. 16.75 × 3.25 ___ 16 × 3
19. 9.1 × 5.4 ___ 8 × 5
20. 8 × 9 ___ 7.5 × 8.25
21. 2 × 4 ___ 2.3 × 4.7

Ring the answer. In each case, only one answer is correct. Decide which is correct without using pencil and paper or a calculator.

22. 4.9 × 0.81   a. 3.969   b. 5.138   c. 49.81
23. 12.4 × 12.4   a. 144.8   b. 53.76   c. 153.76
24. 2.5 × 0.35   a. 8.75   b. 2.85   c. 0.875
25. 4.05 × 4.05   a. 8.4025   b. 16.4025   c. 16.25

LESSON 11.2
Multiplying Two Decimals

Multiply. Check to see that your answers make sense.

① 7
× 4

② 70
× 40

③ 0.07
× 0.4

④ 0.007
× 0.004

⑤ 0.7
× 4

⑥ 19
× 7

⑦ 0.019
× 0.07

⑧ 190
× 17

⑨ 0.0019
× 0.17

⑩ 1900
× 1.7

⑪ 222
× 11

⑫ 2.22
× 1.1

⑬ 0.00222
× 0.011

⑭ 788
× 202

⑮ 7.88
× 2.02

⑯ 78.8
× 2.02

Solve these problems.

⑰ One gallon of gasoline costs $2.25. How much will 8.4 gallons of gasoline cost? _____

⑱ Meredith can jog 6.05 miles in 1 hour. How many miles can she jog in 1.2 hours? _____

⑲ One inch is 2.54 centimeters. Jasmine's puppy grew to 27.25 inches tall. How many centimeters tall is the puppy? _____

⑳ The Corner Coffee Shop sells 1 cup of coffee for $1.75. If Steve buys 3 cups of coffee, how much will it cost him? _____

122 Chapter 11 • *Rational Number and Percent Applications*   **Real Math** • Grade 5 • *Practice*

LESSON 11.3

Name _____ Date _____

Percent and Fraction Benchmarks

Write each percent as a decimal.

① 5% = _____   ② 9% = _____   ③ 59% = _____

④ 95% = _____   ⑤ 23% = _____   ⑥ 32% = _____

⑦ 0.57% = _____ ⑧ 100% = _____   ⑨ 0.7% = _____

⑩ 500% = _____   ⑪ 7.5% = _____   ⑫ 2.3% = _____

Write each decimal as a percent.

⑬ 0.03 = _____   ⑭ 0.09 = _____   ⑮ 3.21 = _____

⑯ 7.3 = _____   ⑰ 1.32 = _____   ⑱ 0.0073 = _____

⑲ 0.416 = _____ ⑳ 0.073 = _____ ㉑ 0.41 = _____

㉒ 8.0 = _____   ㉓ 0.4 = _____   ㉔ 0.088 = _____

Calculate.

㉕ 5% sales tax on $7 _____   ㉘ 8% sales tax on $80 _____

㉖ 3% sales tax on $10 _____   ㉙ 5% sales tax on $5 _____

㉗ 6% sales tax on $30 _____   ㉚ 7% sales tax on $21 _____

㉛ 5% of 100 _____   ㉜ 5% of 200 _____   ㉝ 25% of 200 _____

㉞ 15% of 140 _____   ㉟ 10% of 140 _____   ㊱ 15% of 80 _____

Real Math • Grade 5 • *Practice*   Chapter 11 • *Rational Number and Percent Applications* **123**

Computing Percent Discounts

LESSON 11.4

Calculate the sale price of these items at the Even Steven Store.

	Sale Item	Regular Price	Sale Reduction	Sale Price
1	sewing machine	$289.00	10%	$_____
2	dictionary	$13.98	20%	$_____
3	telescope	$89.98	20%	$_____
4	camera	$354.49	25%	$_____
5	canoe	$175.00	20%	$_____
6	raft	$44.99	20%	$_____
7	mixer	$89.00	$\frac{1}{2}$ price	$_____
8	blender	$39.98	40%	$_____
9	dishwasher	$399.99	$40 off	$_____
10	refrigerator	$449.99	$50 off	$_____

Estimate the amount of a 15 percent tip. Then add the total bill to the tip, and round the answer to the nearest dollar amount.

	Amount	Estimated Tip	Amount Including Tip
11	$24.00	$_____	$_____
12	$40.00	$_____	$_____
13	$15.00	$_____	$_____
14	$50.00	$_____	$_____
15	$18.40	$_____	$_____

LESSON 11.5
Computing Interest

Solve these problems with or without a calculator. Try to approximate your answers in advance. Compare your final answers with those of your classmates.

The National Bank of Mathburg pays 6% interest compounded annually on savings. People who borrow from the bank pay 11% per year on the amount they borrowed.

1 Mrs. Barker put $500 in the bank a year ago. The bank recently paid interest on her $500.

   a. What was the amount of interest? _____

   b. How much does she now have in her account if she has not withdrawn any money? _____

2 Mr. Saxe put $700 in the bank two years ago and has not taken any money from his account.

   a. How much interest did he earn after one year? _____

   b. How much money did he have after one year? _____

   c. How much interest did he earn in the second year? _____

   d. How much money does he have now? _____

3 Mrs. Fumi has borrowed $1,200 from the bank for a year. How much should she pay back at the end of the year? _____

4 The Super Sale Center is having a sale. Every item in the store is to be sold for 20% off the usual price. The store charges a 5% state sales tax. How much would you have to pay for an oak stereo cabinet that costs $350? _____

5 How much would you pay (including tax) for each of these items at the Super Sale Center during the sale? The regular price is given.

   a. toaster, $25   _____

   b. lamp, $34.99   _____

   c. magazine rack, $18.00   _____

   d. soap dish, $8.00   _____

   e. pillow, $15.00   _____

Real Math • Grade 5 • *Practice*   Chapter 11 • *Rational Number and Percent Applications* **125**

LESSON 11.6

Name _____ Date _____

Percents Greater than 100%

Solve the following problems.

1 A year ago, Nicole bought a collectible stamp for $8.00. Now its value is $14.40. Which of the following statements is true? Explain.

   a. The value of the stamp increased by $6.40.

   b. The value of the stamp increased by 80%.

   c. The value of the stamp increased 1.8 times.

   _____

2 Yesterday morning, Ming saw 35 birds eating at the feeders in her backyard. Yesterday evening, she saw 28. Which of the following statements is true? Explain.

   a. The number of birds in the evening decreased by 7.

   b. The number of birds in the evening decreased by 20%.

   c. There were $\frac{1}{5}$ fewer birds in the evening.

   _____

3 Yesterday, the water in Jermaine's swimming pool was 60 inches deep. Overnight there was a heavy thunderstorm, and the water level increased 5%. How deep is the water in the swimming pool now? _____

4 Phil had a puppy that weighed 4.8 pounds when he first bought him. Six months later, the puppy had grown to 14.4 pounds.

   a. How much did the puppy's weight increase? _____

   b. What was the percent increase? _____

   c. After another six months, the puppy had grown to 25.2 pounds. What was the percent increase of its weight over the whole year? _____

5 In March David bought a valuable baseball card for $16.00.

   a. By May the card's value was $22.40. Describe in two statements how the value of the card changed from March to May.

   _____

   _____

   b. In July the value of the card went back down to $17.92. How did the value of the card change from May to July? Describe the change as a percent.

   _____

LESSON 11.7

Name _____ **Date** _____

Probability and Percent

Write the following fractions as percents. If necessary, use a fraction in the percent.

❶ $\frac{1}{5}$ _____   ❷ $\frac{3}{8}$ _____   ❸ $\frac{3}{3}$ _____

❹ $\frac{4}{5}$ _____   ❺ $\frac{2}{3}$ _____   ❻ $\frac{5}{6}$ _____

Complete the following exercises.

Suppose you roll two 0–5 **Number Cubes**.

❼ What are the possible sums that could result?
_____

❽ How many different ways are there to get a sum of 8 with the two cubes? _____

❾ In the table below, list all the possible ways to get each sum from 0 to 10.

Red Cube

	0	1	2	3	4	5
0	0	1				
1		2				
2			4			
3				5	7	
4				7		
5					9	

Green Cube

❿ Use the table to decide what fraction and percent of the time each of the sums below should occur.

a. 0 _____   b. 1 _____   c. 2 _____   d. 3 _____

e. 4 _____   f. 5 _____   g. 6 _____   h. 7 _____

i. 8 _____   j. 9 _____   k. 10 _____

Real Math • Grade 5 • *Practice*   Chapter 11 • *Rational Number and Percent Applications* **127**

LESSON 11.8
Simplifying Decimal Division

Change each of these to an equivalent division problem with a whole-number divisor. Do not do the division.

1. 0.4)9.16 _____
2. 0.867 ÷ 0.59 _____
3. 91.6 ÷ 0.04 _____
4. 0.3)85.5 _____
5. 23.3 ÷ 0.07 _____
6. 0.7)2.33 _____
7. 0.81)9.16 _____
8. .006)3.26 _____
9. 0.15)0.4 _____
10. 40 ÷ 1.05 _____

Ring the appropriate approximation of the three provided.

11. 7.85 ÷ .785   a. 0.1   b. 10   c. 100
12. 3.6)10.66   a. 2.96   b. 0.296   c. 296
13. 0.71)35.78   a. 50.39   b. 5.039   c. 503.9
14. 0.987 ÷ 0.23   a. 0.429   b. 42.9   c. 4.29
15. 568.7 ÷ 49.1   a. 115.8   b. 11.58   c. 1,158
16. 8.2)42.978   a. 5.24   b. 0.524   c. 524

Approximate the answer to each exercise by simplifying each divisor to one digit.

17. 900)8950 _____
18. 4.5)35.1 _____
19. 9.75 ÷ 30.2 _____
20. 67.3)150 _____
21. 81.3 ÷ 409.7 _____
22. 283 ÷ 45.3 _____
23. 91)702.38 _____
24. 4135)8776.8 _____
25. 3,851 ÷ 7,103 _____
26. 0.029)16.85 _____

128 Chapter 11 • *Rational Number and Percent Applications*   **Real Math** • Grade 5 • *Practice*

LESSON 11.9

Dividing Two Decimals

Divide.

1. $0.6 \overline{)4.2}$

2. $56 \div 0.8 =$ _____

3. $0.9 \overline{)450}$

4. $0.08 \overline{)0.24}$

5. $0.6 \overline{)0.42}$

6. $0.8 \overline{)0.56}$

7. $0.45 \div 0.09 =$ _____

8. $0.8 \overline{)0.24}$

9. $0.40 \div 0.05 =$ _____

10. $0.56 \div 0.08 =$ _____

11. $0.9 \overline{)0.45}$

12. $0.08 \overline{)0.024}$

13. $0.9 \overline{)8.1}$

14. $3.2 \div 0.4 =$ _____

15. $0.4 \overline{)3.2}$

16. $2.1 \div 0.3 =$ _____

17. $8.1 \div 900 =$ _____

18. $0.4 \overline{)0.32}$

19. $0.03 \overline{)0.33}$

20. $2.1 \div 0.03 =$ _____

21. $0.81 \div 0.09 =$ _____

22. $0.32 \div 0.008 =$ _____

23. $0.003 \overline{)3.3}$

24. $0.003 \overline{)2.1}$

25. $0.7 \overline{)4.9}$

26. $5.4 \div 0.9 =$ _____

27. $0.08 \overline{)3.20}$

LESSON 12.1

Name _____ Date _____

Estimating Length

Use your hand span as a measuring tool. Stretch out your hand on a ruler, and write down how many centimeters your hand can stretch. Then measure each object by counting the number of times your hand span covers it.

1 Use your hand span to find each measurement.

    a. width of the desktop _____

    b. height of the desk _____

    c. width of your chair _____

2 Compare your results with several classmates' results. Did anyone's answers match yours?

_____

3 Check your measurements by using a ruler, meter stick, or tape measure. If your hand-span measurements were not close, try again.

4 Use the distance between the tip of your thumb and the knuckle to find each measurement in thumb lengths.

    a. thickness of your textbook _____

    b. thickness of the desktop _____

    c. width of your calculator _____

5 Compare your results with several classmates' results. Did anyone's answers match yours?

_____

6 Check your measurements by using a ruler, meter stick, or tape measure. If your thumb-length measurements were not close, try again.

7 Write whether each object would best be measured using thumb lengths, hand spans, or arm spans.

    a. length of a toad _____   **d.** length of a football _____

    b. height of a flagpole _____   **e.** height of a dog _____

    c. height of a giraffe _____   **f.** length of a piece of chalk _____

LESSON 12.2

Name _____ Date _____

Estimating Angles and Distances

Complete these estimating exercises.

1 Estimate the height of the chalkboard in your classroom. Use fists, paces, and a drawing.

   a. Pace off 5 meters from the chalkboard, and then use the fist method to estimate the angle from your eye level to the top of the chalkboard.

   _____

   b. Pace off 10 meters from the chalkboard, and then use the fist method to estimate the angle to the top of the chalkboard.

   _____

   c. Create a scale drawing of your triangle on paper to find the height of the chalkboard. (Don't forget to add the height to your eye.)

   _____

2 Were your estimates in Exercises 1a and 1b the same?

   _____

3 See how close you were in estimating the height of the chalkboard. Measure the height of the chalkboard by using a meter stick or tape measure. Compare this measurement to your estimates. Were they the same?

   _____

4 Use the pace, fist, and a drawing method to estimate the height of your school.

   _____

5 Use the pace, fist, and a drawing method to estimate the height of your school's flagpole.

   _____

6 Find the actual measurements. Compare your estimates to the actual measurements. Were they the same?

   _____

Real Math • Grade 5 • *Practice*

LESSON 12.3

Name _____ Date _____

Applying Customary Measures

Answer each question.

① One can of soda contains 12 fluid ounces. How many fluid ounces are in 6 cans of soda? How many cups is that? _____

② Riley lives on a farm with two horses, Dasher and Pete. Dasher weighs 1,350 pounds, and Pete weighs 1,150 pounds. How many pounds do Dasher and Pete weigh altogether? What is that weight in tons? _____

③ Riley's father uses his truck to take their animals to shows. The truck's hold can carry a maximum of 1.5 tons. Can Dasher and Pete both ride in the truck? Can they ride in the truck with the neighbor's horse, Boomer, who weighs 1,200 pounds? Explain.

_____

Fill in the table to show how much of each ingredient Alex needs to make brownies for a graduation party. His recipe yields 15 brownies, but he needs to make 4 times that amount.

④

Ingredients for 15 Brownies	Ingredients for _____ Brownies
$1\frac{1}{2}$ cups flour	_____ cups flour
$\frac{1}{2}$ cup chocolate chips	_____ cups chocolate chips
2 eggs	_____ eggs
$\frac{3}{4}$ cup sugar	_____ cups sugar
$\frac{2}{3}$ cup pecans	_____ cups pecans
1 teaspoon vanilla	_____ teaspoons vanilla

⑤ The teaspoon from Alex's set of measuring spoons is missing, but he found the tablespoon. Explain how Alex can measure the necessary amount of vanilla by using only the tablespoon.

_____
_____

LESSON 12.4

Name _____ Date _____

Converting Measures

Estimate the equivalents for the following measurements.

1 Darryl is a long-distance runner. Two weeks ago, he ran in a 15-kilometer race. About how many miles was the race?

_____

2 Last summer, Darryl ran in a marathon. A marathon is slightly more than 26 miles. About how many kilometers is that?

_____

3 Scientists from Canada found a dinosaur bone that weighed 12 kilograms. About how many pounds was the dinosaur bone?

_____

4 Natalie is writing a letter to some friends from Israel. She wants to tell them about the dimensions of a baseball field, but first she needs to convert the measurements into metric units.

   a. How many meters are in the 90 feet between home plate and first base? _____

   b. How many meters are in the 127 feet between first base and third base? _____

   c. The pitcher's mound is $10\frac{1}{2}$ inches high. What is its height in centimeters? _____

5 The capacity of the gas tank in Steve's car is 83 liters. What is that capacity in gallons? If gas costs $2.50 per gallon, how much will it cost for Steve to fill his tank?

_____

6 Devon's pen pal from Finland said that his height is 170 centimeters and his weight is 76 kilograms. What is his friend's height in inches? What is his weight in pounds?

_____

Real Math • Grade 5 • *Practice*   Chapter 12 • *Measurement and Graphing* **133**

Lesson 12.5

Measuring Time

Write the number of minutes.

1. $2\frac{1}{3}$ hours = _____ minutes
2. $4\frac{1}{4}$ hours = _____ minutes
3. $1\frac{1}{5}$ hours = _____ minutes
4. $3\frac{1}{2}$ hours = _____ minutes
5. $\frac{3}{5}$ hour = _____ minutes
6. $1\frac{2}{3}$ hours = _____ minutes
7. $2\frac{3}{4}$ hours = _____ minutes
8. 3 hours = _____ minutes
9. $6\frac{2}{5}$ hours = _____ minutes

Write the time indicated in each problem.

10. $2\frac{1}{2}$ hours after 5:20 P.M. _____
11. $5\frac{1}{4}$ hours after 1:20 P.M. _____
12. $3\frac{1}{3}$ hours after 10:15 P.M. _____
13. $\frac{3}{4}$ hour before 9:05 A.M. _____
14. $1\frac{1}{4}$ hours before 8:20 A.M. _____
15. $1\frac{3}{4}$ hours before 4:25 P.M. _____
16. $2\frac{3}{4}$ hours before 1:00 A.M. _____
17. $11\frac{1}{4}$ hours after 8:15 P.M. _____
18. $12\frac{1}{3}$ hours after 1:50 A.M. _____
19. $12\frac{1}{2}$ hours before 8:25 A.M. _____

Solve each problem.

20. David went to a movie that began at 7:15 P.M. When he left the theater, it was 9:38 P.M. How many minutes long was the movie?

21. Mr. Verdi left work at 5:45 P.M. Because of heavy traffic, it took him an hour and a half to get home. What time was it when he got home?

22. Ana went to a play that began at 3:25 P.M. There were two 15-minute intermissions between the acts. The play ended at 6:25 P.M. How many hours did the play last?

LESSON 12.6

Name _____ Date _____

Measuring Circles and Angles

Write how many degrees the minute hand of the clock moves in the given amount of time if the minute hand moves 360° in 1 hour and 6° in 1 minute.

① 27 minutes _____

② $1\frac{1}{3}$ hours _____

③ 8 minutes _____

④ $\frac{1}{6}$ hour _____

⑤ $2\frac{1}{2}$ hours _____

⑥ $1\frac{1}{5}$ hours _____

⑦ $\frac{2}{3}$ hour _____

⑧ 83 minutes _____

Answer these questions.

⑨ If you divide a circle into 5 equally-sized sectors, how many degrees does each central angle have?

_____

⑩ Darius divided a pizza evenly down the middle so he could put different toppings on each half. He put pepperoni on one half and cut that half into 8 equal slices. Then he put sausage on the other half and cut that half into 5 equal slices. How many degrees did each slice with pepperoni have? How many degrees did each slice with sausage have?

_____

⑪ How many degrees does the minute hand on a clock move in 4 hours? How many degrees does the hour hand move in 4 hours?

_____

⑫ How long does it take for the second hand of a clock to move 378°? Explain.

_____

_____

Real Math • Grade 5 • *Practice*   Chapter 12 • *Measurement and Graphing* **135**

LESSON 12.7

Name _____ Date _____

Pictographs and Data Collection

Create a stem-and-leaf plot, and then answer the questions.

1 Jeremy collected the following data about the heights, in inches, of his classmates:
49, 47, 58, 62, 72, 69, 65, 67, 55, 57, 52, 46, 62, 63, 60, 61, 62, 70, 59.

   a. Use these numbers to create a stem-and-leaf plot.

   b. What is the median of the data? _____

   c. Would it make sense to make a pictograph to display this data? Explain.
   _____

Study the pictograph below before answering the questions.

Students' Favorite Classes

Class	Number of Students
English	○ ○ ○ ◐
Geography	○ ◐
History	○ ○ ○ ○
Math	○ ○ ○ ○ ○
Science	○ ○ ○

○ = 4 students

2 By the categories for English and Geography, there are icons that are only partially shown. What do these represent? _____

3 How many students said English was their favorite class? _____
How many said History? _____

4 Which subject received twice as many votes as Geography? _____

5 Which subject was the most popular? How can you tell without counting?
_____

6 Would it be sensible to make a stem-and-leaf plot of the data shown in this pictograph? Explain.
_____

136 Chapter 12 • *Measurement and Graphing*   **Real Math** • Grade 5 • *Practice*

LESSON 12.8

Name _____ Date _____

Making Circle Graphs

Answer each question.

1. Adrianne asked 50 classmates about how they get to school each day, and 17 said they take the bus. How many degrees would this portion have on a circle graph? How many degrees would represent a single response?

   _____

2. If Adrianne increases the scope of her survey to include 150 classmates, how many degrees on a circle graph would represent a single response?

   _____

3. The following table shows the complete results of Adrianne's survey. Make a circle graph to represent this data.

Method of Transportation	Number of Students
Bus	70
Bicycle	26
Carpool	41
Walk	13

4. How many degrees did you use to represent each portion of your circle graph? Explain how you found each angle.

   _____
   _____
   _____

5. There are a total of 634 students at Adrianne's school. How could you use the survey to guess how many of them ride the bus to school? Make an estimate to the nearest hundred.

   _____
   _____

LESSON 12.9

Name _____ Date _____

Creating and Using Graphs

Make a bar graph using the table below, which shows the number of cans collected by each grade for a school's canned food drive. Then answer the questions.

Grade	K	1	2	3	4	5	6
Number of Cans	78	54	93	65	63	80	72

❶ Which grade collected the most cans? _____

❷ Which grade collected $\frac{3}{4}$ as many cans as the sixth grade? _____

❸ Which grade collected 15 more cans than the third grade? _____

Use the graph to answer the questions.

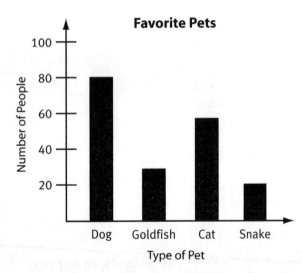

❹ Which animal was named by the most people? _____

❺ Which animal was named by the fewest people? _____

❻ How many more people chose goldfish than snakes? _____

❼ Altogether, how many people were asked to name their favorite type of pet? _____

LESSON 12.10

Name _____ **Date** _____

Making Line Graphs

Make a single line graph from the data in the table below, which shows the populations of two counties, in thousands of people, over fifty years. Then use the graph to answer each question.

Year	1950	1960	1970	1980	1990	2000
Thurn County	46	70	97	147	186	218
Taxis County	133	142	149	145	139	134

1 How have the populations of each county changed overall?

_____

_____

2 When did the population of Thurn County surpass the population of Taxis County?

_____

3 When was the difference between the two populations the greatest?

_____

4 About when was the population of Taxis County the greatest?

_____

5 Between what years did the population of Thurn County increase the most?

_____

Make a single line graph from the data in the table below, which shows the weights, in ounces, of three puppies at the beginning of weeks 1–5. Then use the graph to answer each question.

Week	1	2	3	4	5
Stubb	13	19	30	45	60
Billy	15	22	35	42	55
Pip	8	19	28	30	49

6 Pip was sick during one week and did not grow as much as usual. What week was this? _____

7 When were Stubb and Pip the same weight? _____

8 Which puppy weighed the most at the beginning of the five weeks? Which puppy weighed the most at the end? _____

Real Math • Grade 5 • *Practice*   Chapter 12 • *Measurement and Graphing* **139**

LESSON 12.11

Name _____ Date _____

Interpreting Graphs

Use each graph to answer the questions.

1. The graph shows a town's percentage expenditures for parks and recreation and for street repairs and improvements for four years.

   a. In what year was about 5% spent on improvements to parks? _____

   b. In what year was about 6% spent on improvements to streets? _____

   c. Describe the greatest increase in spending from one year to the next.

   _____

   d. In which year(s) was the spending for parks greater than the spending for streets?

   _____

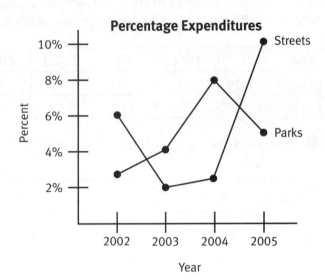

2. In a local junior high school, the number of boys and girls who attend grades 6, 7, and 8 are displayed in the bar graph.

   a. In which grade are there more boys than girls?

   _____

   b. In which grade is there the greatest difference in the number of boys and girls?

   _____

   c. How many boys are in the seventh grade? How many girls?

   _____

   d. In which grade are there the most students altogether?

   _____

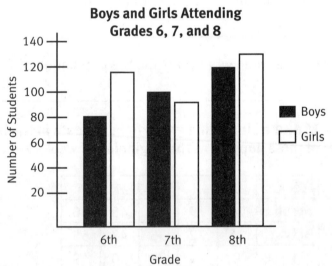

140 Chapter 12 • *Measurement and Graphing*   Real Math • Grade 5 • *Practice*

Name _____ Date _____

Centimeter Graph Paper

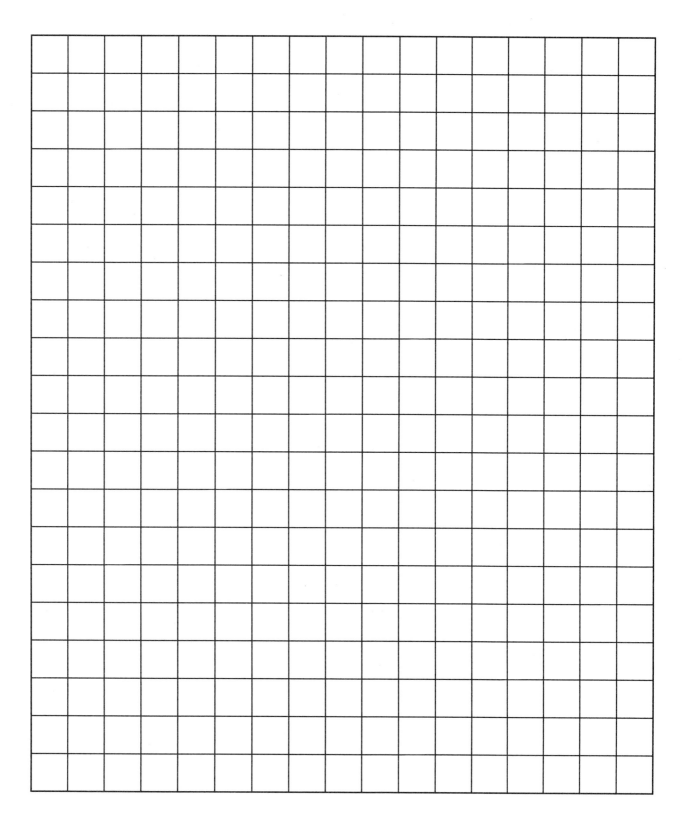

Real Math • Grade 5 • *Practice* Masters • *Centimeter Graph Paper* **141**

Name _____ Date _____

Graph City

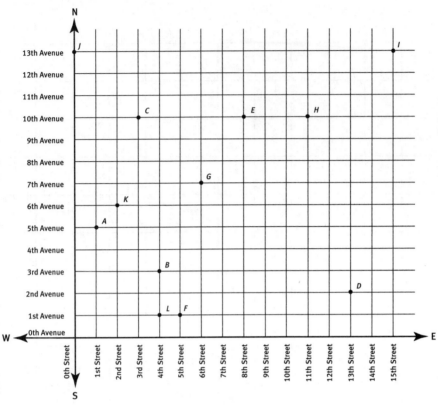

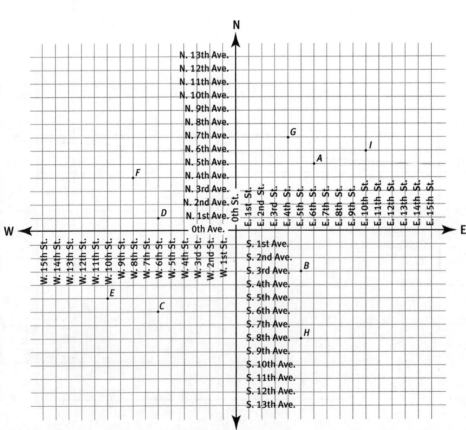

142 Masters • *Graph City*

Real Math • Grade 5 • *Practice*

Name _____ **Date** _____

Decimal Equivalents of Fractions

$\frac{1}{2}$

$\frac{1}{3}$ $\frac{2}{3}$

$\frac{1}{4}$ $\frac{2}{4}$ $\frac{3}{4}$

$\frac{1}{5}$ $\frac{2}{5}$ $\frac{3}{5}$ $\frac{4}{5}$

$\frac{1}{6}$ $\frac{2}{6}$ $\frac{3}{6}$ $\frac{4}{6}$ $\frac{5}{6}$

$\frac{1}{7}$ $\frac{2}{7}$ $\frac{3}{7}$ $\frac{4}{7}$ $\frac{5}{7}$ $\frac{6}{7}$

$\frac{1}{8}$ $\frac{2}{8}$ $\frac{3}{8}$ $\frac{4}{8}$ $\frac{5}{8}$ $\frac{6}{8}$ $\frac{7}{8}$

$\frac{1}{9}$ $\frac{2}{9}$ $\frac{3}{9}$ $\frac{4}{9}$ $\frac{5}{9}$ $\frac{6}{9}$ $\frac{7}{9}$ $\frac{8}{9}$

$\frac{1}{10}$ $\frac{2}{10}$ $\frac{3}{10}$ $\frac{4}{10}$ $\frac{5}{10}$ $\frac{6}{10}$ $\frac{7}{10}$ $\frac{8}{10}$ $\frac{9}{10}$

⟵─┼────┼────┼────┼────┼────┼────┼────┼────┼────┼────┼─⟶
 0 0.1 0.2 0.3 0.4 0.5 0.6 0.7 0.8 0.9   1

Real Math • Grade 5 • *Practice*   Masters • Decimal Equivalents of Fractions **143**

Masters

Name _____ Date _____

Fractured Fractions

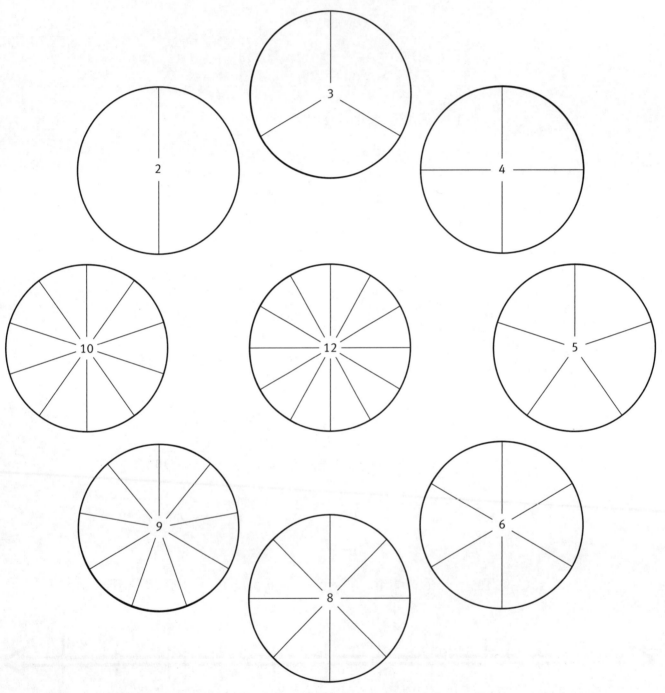

144 Masters • *Fractured Fractions*

Real Math • Grade 5 • *Practice*

Building a Flexagon

1 Mark your strip of ten triangles like this.

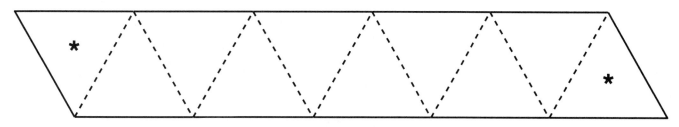

2 Place your strip of triangles on the guide below, and fold as shown.

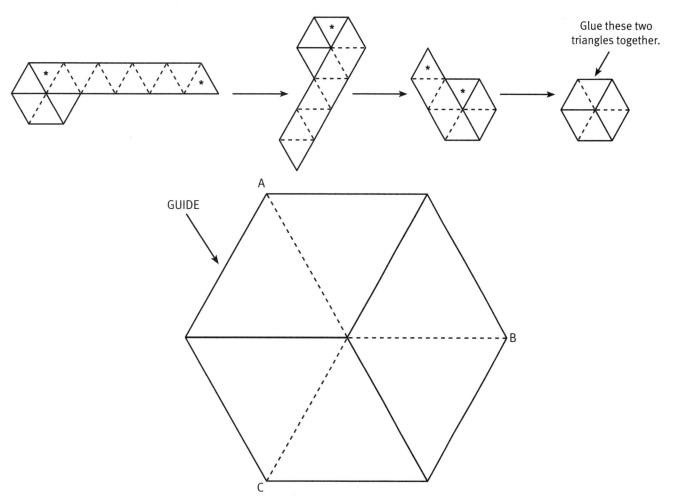

3 Glue the two starred triangles together. When using the gummed tape, one of two things will happen to the overlapping triangles at the top. Either the glossy faces are already touching each other (in which case, you can moisten one and glue these together), or the dull faces are touching each other. If this occurs, lift the bottom triangle and let the top triangle slide beneath it so the glossy faces are touching. Then moisten one triangle, and glue these together.

Real Math • Grade 5 • *Practice*

Graphs

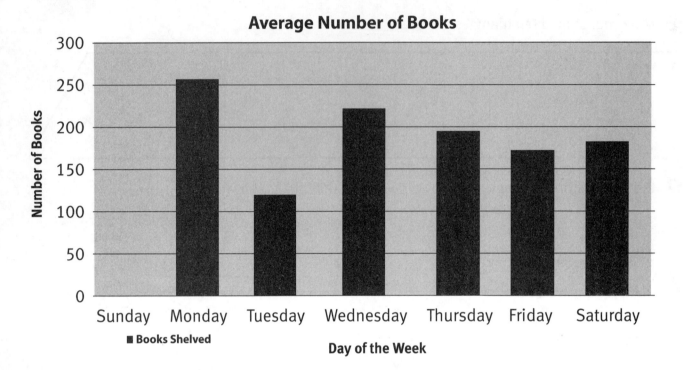

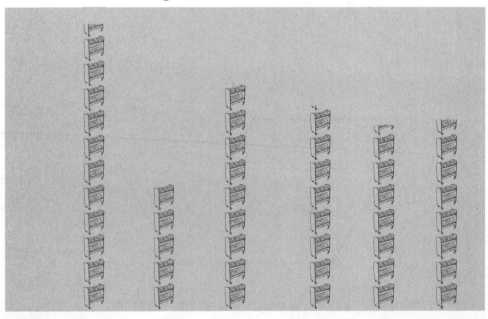